What Your Colleagues Are Saying . . .

If you share my belief that "construct viable arguments and critique the reasoning of others" are perhaps the nine most important words in the Common Core era, then *Mathematical Argumentation in Middle School* is just what you need. This powerful and practical book takes us through an accessible process of generating cases, making conjectures, and justifying that fully supports bringing SMP #3 to life in our classrooms.

—Steve Leinwand
American Institutes for Research

This great resource gives teachers tools to implement the four cycles of mathematical argumentation and help students develop a "variety of expertise," as described in the Standards of Mathematical Practice. As students cycle through the phases, they are guided in building "mathematical authority" as independent thinkers and creators of mathematical ideas. I recommend this book to any teacher who wants to amp up the math discussion in their classroom.

—Annette Hilts
High School Mathematics Coordinator
Vallejo City Unified School District

Now more than ever, we need to provide *all* children with opportunities to learn to think critically and participate in thoughtful, productive debate in today's society. This book translates the mathematical practice of argumentation into a four-stage process that can be applied across a wide range of mathematical content. This process utilizes an innovative, research-based approach based on improv games that opens access for all students to participate in the process of mathematical argumentation. Finally, there is a practical guide for making argumentation an everyday practice in mathematics classrooms!

—Kristen Bieda
Associate Professor of Teacher Education
Michigan State University

MATHEMATICAL ARGUMENTATION

IN MIDDLE SCHOOL—
THE WHAT, WHY, AND HOW

A Step-by-Step Guide With Activities, Games, and Lesson Planning Tools

Jennifer Knudsen

Harriette S. Stevens

Teresa Lara-Meloy

Hee-Joon Kim

Nicole Shechtman

CM CORWIN MATHEMATICS

FOR INFORMATION:

Corwin

A SAGE Company

2455 Teller Road

Thousand Oaks, California 91320

(800) 233-9936

www.corwin.com

SAGE Publications Ltd.

1 Oliver's Yard

55 City Road

London, EC1Y 1SP

United Kingdom

SAGE Publications India Pvt. Ltd.

B 1/I 1 Mohan Cooperative Industrial Area

Mathura Road, New Delhi 110 044

India

SAGE Publications Asia-Pacific Pte. Ltd.

3 Church Street

#10-04 Samsung Hub

Singapore 049483

Program Manager, Mathematics: Erin Null

Editorial Development Manager: Julie Nemer

Editorial Assistant: Nicole Shade

Production Editor: Melanie Birdsall

Copy Editor: Liann Lech

Typesetter: Integra

Proofreader: Wendy Jo Dymond

Cover Designer: Scott Van Atta

Marketing Manager: Margaret O'Connor

This material is based upon work supported by the National Science Foundation under Grants No. DRL-14178950, DRL-0455868, and DRL-1119518. Any opinions, findings, and conclusions or recommendations expressed in this material are those of the authors and do not necessarily reflect the views of the National Science Foundation.

Use of GeoGebra software by purchasers of this book shall be subject to GeoGebra's terms of use.

Use of Desmos software by purchasers of this book shall be subject to Desmos's terms of use.

Printed in the United States of America

ISBN 978-1-5063-7669-1

This book is printed on acid-free paper.

Certified Chain of Custody

SUSTAINABLE FORESTRY INITIATIVE

Promoting Sustainable Forestry

www.sfiprogram.org

SFI-01268

SFI label applies to text stock

17 18 19 20 21 10 9 8 7 6 5 4 3 2 1

Contents

Visit the companion website at

resources.corwin.com/mathargumentation

for a Professional Development Guide and Argumentation Lesson Planning
Template, as well as posters, handouts, digital mathematics tools,
and other resources.

Preface

WHY THIS BOOK NOW?

This book is a comprehensive guide to fostering and supporting mathematical argumentation among your middle school students. We offer it to all teachers who are currently grappling with how to increase and strengthen mathematical discourse in their classrooms. It is designed to be a practical guidebook, with advice on teaching, activity handouts, links to interactive web pages that go with the activities, warm-up games, and lesson planning ideas.

Mathematical argumentation is important not only because it appears in current state standards but, more fundamentally, because it is also what mathematicians do. All students deserve to have access to this high-level disciplinary practice. Additionally, the principles and logic of mathematical argumentation are very relevant to the kinds of reasoning your students need to be involved citizens and to have successful careers.

Experience with mathematical argumentation is critical for students in middle grades. For too long, argumentation has been associated with two-column geometric proofs, taught by rote in high school. While numerous studies have documented high school and even college students' troubles with proof (e.g., Healy & Hoyles, 2000), there are also promising studies of even elementary school students learning to make generalized arguments (e.g., Ball & Bass, 2000; Schifter, 2009). Middle school is a crucial transition period when students refine their understanding of generalization by paying attention to patterns and move beyond the consideration of individual cases.

All teaching is complex work, and teaching for argumentation is even more so. We hope to honor your complex work by portraying it and guiding it as "disciplined improvisation" (Sawyer, 2011). We help you develop specific teaching moves you can call on when you have to think on your feet as a surprising argument unfolds in class. Teaching moves are behaviors that are aimed at a purpose: For example, we provide both specific questions you can ask and the reasons for asking them.

HOW THIS BOOK WORKS

The book has eight chapters.

Chapter 1 is an introduction to mathematical argumentation and why you should teach it.

Chapters 2, 3, 4, and 7 are each devoted to one part of our model of argumentation—generating cases, conjecturing, justifying, and concluding.

Justifying is so important that we devote two more chapters to it: Chapter 5 on the use of representations and Chapter 6 about the levels of justifications your students will make.

The final chapter, Chapter 8, is about how to plan to put together all the pieces presented in the earlier chapters.

Most chapters have at least one vignette from the classroom. These vignettes are sometimes based on what we have observed in the classrooms of teachers who were participating in the Bridging Professional Development program that we work on at SRI International, but they also portray a composite of several teachers and our vision for how teachers and students can work together to create arguments. The purpose of the vignettes is to help you develop your own vision of teaching for argumentation, as well as provide concrete examples of the teaching moves you will need to support argumentation in your classroom.

The lead vignettes in Chapters 2, 3, 4, and 7 tell the story of Ms. Cooper and her students working on a series of lessons that lead them through the stages of argumentation about coordinate geometry. Other vignettes portray single lessons that address one or more parts of argumentation. We see both types of lessons as important—those that fit together to cover all the parts of argumentation and those that focus on one part of argumentation. Both types of lessons must address important content goals, and the lessons in our vignettes do so.

Following the vignettes, Chapters 2 through 5 and 7 on the parts of the model each contain four more sections:

- **Teaching moves** that can be used to elicit cases, conjectures, justifications, or conclusions from students
- **Classroom norms** that are important for each part of argumentation and warm-up games that you can use to help inculcate these norms
- **Planning** for each part of argumentation that will help you think through in advance how to deploy teaching moves and select tasks
- **Mathematical tasks** appropriate for the middle grades, covering important content aligned with many current state standards. Tasks are accompanied by activity handouts structuring the relevant parts of the model and, where appropriate, interactive web pages created with GeoGebra and Desmos, where students can investigate geometric shapes and algebraic representations.

Chapter 6 presents different levels of sophistication of students' justifications. This chapter is based on classic research in this field and supplies teaching moves for helping students increase the sophistication of their arguments.

Chapter 8 provides tools and a process for planning argumentation-rich lessons for use with your students. A *planning form* guides you in thinking about which particular moves may be useful in that lesson, as well as possible student responses. Our *visualization technique* helps you turn those written plans into rich imaginings of what may happen that will guide you when you actually teach a lesson.

Every chapter ends with a Working Together section with an agenda for working with other teachers in a professional learning community to implement the advice given in each chapter.

Reaching the full diversity of students is central in our work. In Access for All sections, found in shaded boxes throughout the book and marked with an icon, we highlight how our methods address the needs of students who have been historically marginalized in mathematics—students of color, students with learning differences, and students who are English Language Learners (ELLs). Often, digital tools provide means for access for all, and there is a section on digital tools in almost every chapter.

ACCESS FOR ALL

HOW YOU CAN USE THIS BOOK

You don't have to read the whole book to get started. In Chapter 1, we offer four steps for getting started with argumentation that you can begin tomorrow. You can use justification in almost any mathematics lesson you normally teach, as you get students to discuss the reasons behind the answers they give to problems. Then, as you get further into the book, you can focus on, for example, promoting bold conjectures as you read the chapter on conjecturing. Each chapter contains advice you can implement as you gain confidence in teaching for argumentation. If you are already integrating argumentation into your mathematics discourse, you can choose chapters on areas in which you want additional help.

HOW WE DEVELOPED THE APPROACH IN THIS BOOK

The advice in this book was derived from our work over the past 8 years on Bridging Professional Development, a project funded by the National Science Foundation and conducted through SRI International. Bridging helps teachers learn specific teaching moves for supporting mathematical argumentation through specially designed curricula and through the use of teaching games based on improvisational theater games. These games also inspire activities for establishing productive norms in the classroom—warm-up games—as we describe across several chapters in the book.

In three Bridging projects, we worked with teachers in a diversity of districts through several iterations of our professional development program. Much of our program design and approach to argumentation is rooted in on our team's collective experiences in the classroom and as teacher professional developers. We also drew from the best research on argumentation done over the past 3 decades. We are particularly indebted, for example, to the work of Imre Lakatos for sparking some of our initial thinking about the nature of mathematical argumentation. Through each iteration of our program, we worked closely with teachers in co-design activities and other rich interactions to refine our approach and explore in depth the kinds of supports teachers would really need to broaden access for a diversity of student populations to mathematical argumentation.

We also had promising results in two small-scale studies with teachers participating in Bridging programs. In one study, we found that providing teachers with curriculum activities focused on conjecturing and justifying, along with training to use teaching moves to support argumentation, enabled them to facilitate argumentation talk in their classrooms. In another study, we provided teachers with a 2-week unit focused on learning geometry through argumentation, and we found that students demonstrated learning gains from pretest to posttest. While these findings are promising in general, you will need to use your own judgment as a practitioner to decide how to apply Bridging's approach in your own unique setting. You can read more about the details of our professional development program and research in these additional publications: Knudsen, Lara-Meloy, Stevens, and Rutstein, 2014; Knudsen and Shechtman, 2017; Knudsen, Shechtman, Lara-Meloy, Stevens, and Rutstein, 2015; and Shechtman and Knudsen, 2011.

Acknowledgments

We wish to thank a number of people and organizations:

- The Bridging team, which has included Kate Borelli, Ricky Carter, Claire Christensen, Susan Empson, Bowyee Gong, Amy Hafter, Joanne Hawkins, Danae Kamdar, Yesica Lopez, Vera Michalchik, Savitha Moorthy, Blanca Olmos, Deepa Patel, Aliya Pilchen, Ken Rafanan, David Reider, Daisy Rutstein, Maria Salciccioli, Gucci Trinidad, Phil Vahey, and Brenda Waller, for their commitment and creativity
- The Center for Technology in Learning at SRI International, and its co-directors Jeremy Roschelle and Barbara Means, who supported this work with their encouragement and the gift of time
- Carrie Lobman and Richard Cox for helping us understand applied improv for the classroom and for teacher professional development
- All the teachers who have been co-designers with us and participants in Bridging Professional Development, who welcomed us into their classrooms, and who provided us with both feedback on the book and the quotes contained within it. We learned as much from them as we hope they did from us.

We each have individual thanks to give as well:

Jennifer Knudsen

I am deeply grateful to my husband, Dale Owen Crouse, for his unfailing love, support, and patience—and intensive editorial advice. I thank my daughter, Xenia Dawn Crouse Knudsen, for her inspiring enthusiasm for mathematics. I thank them both for keeping the household running over the past year while I was holed up writing. My sisters, Wendy, Karin, and Glenda; my mother-in-law, Betty; and my dear friend, Susan, provided love and encouragement, and I am grateful to each of them. I greatly appreciate my coauthors and our collaborative process, developed over many years of working together.

Harriette S. Stevens

I am grateful to my husband, Myles Stevens, for his patience and willingness to listen and review numerous drafts of my work, and my children, Michelle Stevens Jones, MD and Myles III, who continued to encourage me. Their perspectives helped me to gain a deeper understanding of what it means to provide better communications in a mathematics classroom.

To my parents, Isaac and Eva Stallworth, who were lifetime educators, and my mentor, Herbert Stallworth—their compassion for learning made me believe that all students can demonstrate their mathematical talents.

Three years ago, Jennifer Knudsen said that we needed to write a book about our work in supporting mathematical argumentation in urban classrooms. A special thanks, Jennifer, you made this book a reality.

Teresa Lara-Meloy

I'm grateful to Jennifer for the opportunity to continue learning mathematics through the Bridging Professional Development project and for inviting me to collaborate on this book. I'm thankful to my parents for their encouragement: my mother, Mary Meloy, for teaching me endless curiosity, and my father, Jorge Lara y Góngora, for instilling in me argumentativeness and answer seeking. Finally, I'm also grateful to my husband and daughters for allowing me time and space to focus on this book collaboration.

Hee-Joon Kim

I want to thank every single teacher who participated in Bridging projects. They are the foundation and inspiration for my work, and I learned the most from their efforts to make a difference in their classrooms. I am grateful to Dr. Susan Empson for giving me the opportunity to become a part of the Bridging Project, work with a great team, and grow as a researcher. I am also thankful to my parents and husband for giving me enduring support and love.

Nicole Shechtman

I am grateful for the passion that our teachers have brought to their own learning and the passion they have for bringing new opportunities to their students. They do challenging work, and their enthusiasm is contagious. I am also grateful to the Bridging team and what we have learned together in our journey. Special thanks to David for providing both love and chocolate all along the way.

PUBLISHER'S ACKNOWLEDGMENTS

Corwin gratefully acknowledges the contributions of the following reviewers:

Rebekah Elliot
Associate Professor
Oregon State University
Corvallis, OR

Annette Hilts
High School Instructional
    Coordinator
Vallejo, CA

Steve Leinwand
Principal Research Analyst
Washington, DC

Liz Marquez
Assessment Developer
Princeton, NJ

Erin Pfaff
Graduate Student
Nashville, TN

Lisa Usher-Staats
Retired LAUSD Mathematics
    Coordinator
Long Beach, CA

Cathy Yenca
Mathematics Teacher
Austin, TX

About the Authors

Jennifer Knudsen has been working in mathematics education since her days as a Peace Corps volunteer in Kenya and as a teacher in New York City Public Schools. She has focused on students' engagement in mathematics as an equity issue throughout her career, including work on numerous curriculum and professional development projects. She directs the Bridging Professional Development project as part of her role as a senior mathematics educator at SRI International. She holds a BA from The Evergreen State College, where she learned to love mathematical argumentation. She lives in Austin, Texas, with her husband and daughter.

Harriette S. Stevens attended the University of Kansas where she received her Bachelor of Arts in applied mathematics and Master of Arts in education, with a concentration in mathematics. She received her doctorate in education, with a focus on curriculum and instructional design, from the University of San Francisco. She was the director of a mathematics professional development program for K–12 teachers at the University of California, Berkeley's Lawrence Hall of Science. In this capacity, she worked in partnership with several urban school districts and designed professional development and instructional materials to help improve teachers' understanding of mathematics content and their students' preparation for success in college and careers. Currently, she is a consultant with the Mathematics Education Group, San Francisco, and co-director of the Bridging Professional Development project at SRI International, Menlo Park. Her interests include a focus on strengthening teachers' knowledge of mathematics content and the ways in which this knowledge is used to advance classroom discourse and problem solving in urban schools.

Teresa Lara-Meloy is passionate about finding better ways of teaching middle school mathematics and improving ways to support teachers. As Math Ed Researcher at SRI International, she designs technology-integrated curricular and professional development materials. She received her MEd from Harvard's Graduate School of Education. She is a member of the NCSM and TODOS. She has co-authored articles on technology in education and the role of technology in supporting the participation of English Language Learners in math class.

Hee-Joon Kim, PhD, is a mathematics education researcher at SRI International located in Menlo Park, CA. Her research focuses on understanding classroom discourse that supports mathematical argumentation in middle school. She has expertise in designing curriculum materials with dynamic tools for students in middle grades. She has been involved in research-based professional development projects that focus on improving classroom practices that support conceptual understanding and promote equity. She received a BS in mathematics at Ewha Womans University in South Korea and a PhD in mathematics education at the University of Texas at Austin.

Nicole Shechtman, PhD, is a senior educational researcher at SRI International located in Menlo Park, CA. Her research and evaluation work explores critical issues in mathematics teaching and learning, innovative uses of educational technology, and the development of social and emotional competencies, such as effective communication, teamwork, and everyday problem-solving. She holds a PhD in psychology from Stanford University.

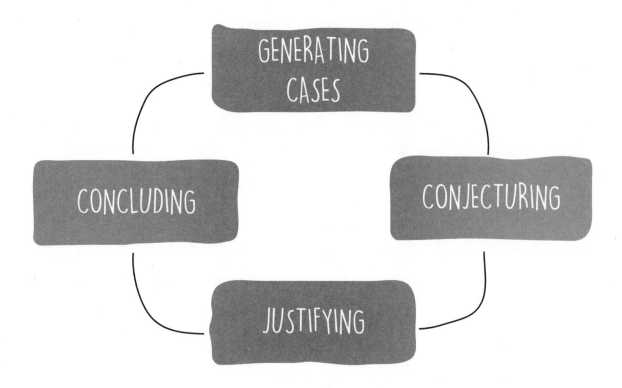

CHAPTER 1

Mathematical Argumentation
Why and What

In this chapter, you will learn

- What argumentation is and what it is not
- How to use a four-part model of argumentation: generating cases, conjecturing, justifying, and concluding
- About argumentation as a social process
- Why teaching is disciplined improvisation and how improvisation supports argumentation, norm setting, and student engagement
- Steps for introducing argumentation in your mathematics classroom
- About argumentation *in* lessons and argumentation lessons
- How to share new ideas for teaching mathematical argumentation in working together with your colleagues

ARGUMENTATION IS IMPORTANT!

> There are things that we need to communicate in everyday life, especially in the society in which we find ourselves now with all kinds of complexities. If we could just step back and think critically about it, then we should be able to come up with some kind of solution to the problems we face. That's why I just think that math argumentation is so great, not only for education, but so that you will be able to function as a human being and a citizen in this society.
> —Seventh-grade mathematics teacher

That's what a middle school mathematics teacher, one of our workshop participants, had to say about the potential for mathematical argumentation to make an impact outside of the classroom. What if students can use the same kind of reasoning to solve problems in their lives as they do to, say, establish that the sum of two odd numbers is an even number? As we consider our students' futures, the kind of careful reasoning that they do together in classroom mathematical argumentation is an important 21st century workplace and life skill. Making logical connections among abstract ideas and interacting with others to clarify their ideas are both deemed necessary in an increasing number of good jobs (Partnership for 21st Century Skills, 2008).

And then there are current mathematics standards. You probably know about the emphasis on mathematical practices or processes in most current state standards, including the practices that students "construct viable arguments and critique the reasoning of others" (National Governors Association Center for Best Practices, Council of Chief State School Officers, 2010) and "justify mathematical ideas and arguments using precise mathematical language in written or oral communication" (Texas Education Agency, 2012). The ability to make sense of the story mathematics tells, construct a viable argument about that story, justify one's reasoning, and critique the reasoning of others are essential skills in almost every line of work and in citizen participation.

In addition to these practical considerations, we believe that the practice of mathematical argumentation is the most important of the mathematical practices because it is the fundamental way in which mathematicians communicate with each other. The search for mathematical truth is ongoing, as mathematicians create new ideas and justify them and as students reason together in a classroom.

Furthermore, access to mathematical argumentation is an equity issue. Every student should have access to this high-level disciplinary practice. Providing this access in elementary and middle school puts students on a path to higher level mathematics in high school and college. Current research indicates that about a third of the difference in mathematics achievement between students of color and white students, and between students from low- and high-income families, is attributable to the *opportunity to learn high-level mathematics* that they are given in class (Schmidt, Burroughs, Zoido, & Houang, 2015). The techniques in this book provide practices that support equitable access to high-level mathematics.

Although argumentation is serious business, it's also true that engaging in argumentation can make your mathematics classroom more joyful. Students get to play with mathematical ideas and take ownership of them in a way that often delights them. You'll most likely feel a boost yourself. One participant in our early workshops proclaimed that every Friday was argumentation day, and her class eagerly looked forward to it. While we advocate including argumentation most days, not just Fridays, we appreciated the spirit of her designation and found it a positive step in her own professional development.

The approach, techniques, and activities in this book were developed while working with teachers in a variety of settings. We have worked, in particular, with teachers in urban schools with high proportions of youth of color and students from low-income families. We have also worked in schools in more affluent communities. Teachers in all of these settings have used these methods to bring argumentation to their classrooms. Additionally, we have worked with teachers of students who receive special education services and students who are English Language Learners, and they have found that their students, too, can participate in mathematical argumentation. The vignettes and examples we present are informed by what we learned from these teachers but do not represent any one teacher.

WHAT ARGUMENTATION IS—AND IS NOT

To understand what mathematical argumentation is, it is important to understand what it is not. We can contrast it with other mathematical practices and processes. Take, for example, a graph of distance as a function of time, as shown in Figure 1.1. It can be the starting place for lessons on problem solving, modeling, or argumentation, depending on the prompt that goes with the graph. A problem-solving prompt, for example, is "Create a trip with three segments that ends at 200 feet." It calls for a solution, carefully reasoned but not necessarily an argument. On the other hand, a prompt that calls for argumentation goes like this: "Raj says that if one line is steeper than another, then it represents a faster motion. Is this always true?" Notice the question, "Is this always true?" In this book, we help you develop a repertoire of ways to use that simple question, among others, to engage your students in building arguments throughout the school year.

FIGURE 1.1 Time Versus Position Graph

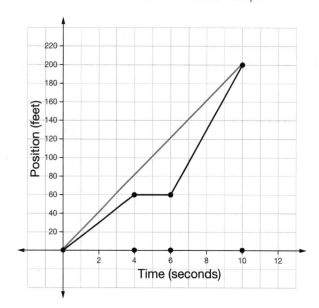

A second question asks for argumentation: "How do we know it is true?" In this question, the focus is on a public demonstration of why a statement is true or false. The onus is on students to come up with an argument that is convincing to others. This press for truth is key in fostering argumentation that takes place *among* students so that students not only construct arguments but also critique each other's reasoning.

This approach to argumentation positions it as a social practice—what we engage in to find out the truth together (Thurston, 1998). For example, you can tell students that the area of a parallelogram is calculated by multiplying the lengths of the base and height. What if students multiply the base, height, and the other side length to find the area? You could simply tell them this is wrong. But it is more powerful for students to explain to each other why multiplying these three numbers together does not make sense, calling on the concept of area as a measure of two-dimensional space.

Students will likely also need help understanding what argumentation is and what it is not. They may bring their own notions of what an argument is—for example, a fight—and it will take some work to help them develop a new way of thinking about argumentation as a mathematical practice, as a way of reasoning together about the truth. We'll have more to say about classroom norms for argumentation in the following chapters, but this may be the most fundamental norm of all: *We are finding out the truth together.*

A FOUR-PART MODEL OF ARGUMENTATION

Our model of argumentation is based on what mathematicians do and what philosophers and educators have posited as parts of argumentation. We distilled the experts' (e.g., Harel & Sowder, 1998, 2007; Krummheuer, 1995; Lakatos, 1976) views on argumentation into a structure that works for teachers and students just beginning with argumentation as well as for those with more experience. The model has four parts:

- Generating cases—creating something to argue about
- Conjecturing—making bold claims
- Justifying—building a chain of reasoning
- Concluding—closure on truth or falsity

In practice, the parts of the model may get mixed together, and the process is often iterative. When you are starting out, it's useful to think of each part as a separate activity (Figure 1.2).

FIGURE 1.2 Four-Part Model for Argumentation

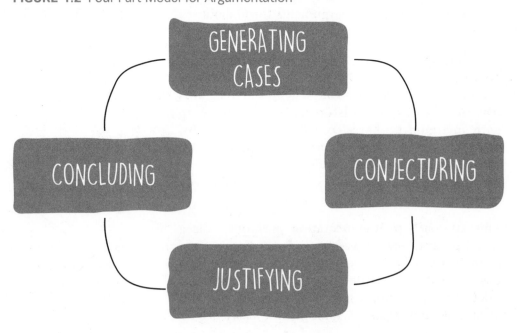

Important to our model is that you can make some teaching moves to elicit students' mathematical work in each part. Already, we've introduced two teaching moves—asking "Is this always true?" and "How do we know it is true?"—that you can use to stimulate mathematical argumentation among your students. Obviously, there's more to it than that. Next, we'll walk through the model for argumentation that you can use so that students get to experience all the important parts of an argument.

Generating Cases—Creating Something to Argue About

In order for students to make a mathematical argument, they need some mathematics to argue about. That may seem pretty obvious, but the challenge is in finding the right tasks. If you have focused on problem-solving lessons or supporting rich discussion in your class, you've probably used activities in which students do what we call "generating cases." The cases often reveal patterns that students are asked to describe and explain. For example, when considering the sums of even and odd numbers—a way to introduce variables—students can explore the results of adding different numbers such as 5 + 8 as an example of odd plus even and 6 + 10 as an example of even plus even. As students come up with examples to try, they are, in the terms of our model, generating cases. In a set of cases, they can observe patterns. Often, lessons that include patterning stop with recognizing the pattern. But argumentation is just beginning as students recognize patterns across cases. Students' descriptions of patterns can be clearly articulated as conjectures, which is the next part of argumentation.

We say more about what makes for good generation of cases in Chapter 2, as well as provide tasks well suited to this purpose.

Conjecturing—Making Bold Claims

Conjectures are mathematical statements that can be determined to be true or false. They often have a level of generality that goes beyond a single case. An important class of general conjectures includes the answers to the question, "What must *always* be true?" But specific conjectures can be important as well. Specific conjectures might be about the solution to a particular equation or an observation about a regular polygon. When specific or general conjectures are appropriate is discussed more in Chapter 3.

> One easy way I can get students to make different conjectures, even in a lesson on procedures, is by asking for consensus on answers to questions, before I give the answer myself.
>
> —Sixth-grade teacher

Keep in mind that conjecturing is a time for students to make bold claims that go beyond the obvious. You don't want the only arguments that your class makes to be about statements that students already know are true; it takes the adventure out of argumentation, as well as some of the purpose. Also, arguing for a conjecture that turns out to be false can be a powerful learning activity, lending more insight into the concept behind the conjecture than might otherwise be accessible to your students. When students are conjecturing, you want to encourage bold claims with questions such as "What do you think *might* be true about all the cases you have looked at so far?"

A *move* is the smallest piece of behavior that can be aimed at a purpose. We discuss teaching moves throughout this book as questions you can ask or actions you can take, along with their specific purposes.

Conjectures are mathematical statements that can be determined to be true or false.

In Chapter 3, we explain ways to elicit bold conjectures, ways to handle multiple conjectures, and ways to start a lesson with a single conjecture you select in advance with a particular purpose in mind.

Justifying—Building a Chain of Reasoning

Justifying occurs when students present reasons for why a conjecture is true or false.

A justification is a logically connected chain of statements that begins with what students already know to be true and ends by establishing the truth or falsity of a conjecture. When students justify a conjecture, they are presenting the reasons why it is true, and those reasons need to form a connection between what is known and what is yet to be known. Don't expect, though, that a justification comes out in perfect order in the social process of classroom argumentation. The logical chain may not be obvious and will have to be articulated in the next phase, concluding.

The simplest justification is a counterexample—a single example that shows that a conjecture is false. For example, after establishing that the sum of two even numbers is even, students may conjecture that

> The sum of two odd numbers must always be odd.

Say a student points out that $3 + 5 = 8$. That one example tells us that the sum is *not*, in fact, always odd, thereby serving as a counterexample, which is a complete justification that the conjecture is false.

A counterexample often suggests the next conjecture: In this case,

> The sum of two odd numbers is always even.

There are several ways to make a mathematically sound justification for this conjecture using different representations. The choice of representations is important in justifying; students need to be somewhat comfortable with the representation they are using, although they don't have to be expert with it. Students can make good justifications using very simple representations. For example, one justification that the sum of two odd numbers is even could rely on diagrams (Figure 1.3).

FIGURE 1.3 Justification Using a Pictorial Representation

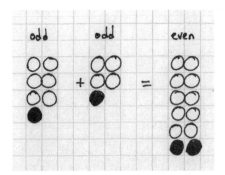

(For more on pictorial representations in justification, see Schifter [2009] and our Chapter 5.)

Each odd number can be represented as a set of paired dots plus one extra dot. When two odd numbers are added together, the result shows all the pairs of dots and the two extra dots that form another pair. Because the number can be represented with pairs of dots only, it is even.

Other students may use variables in their justification. They can represent the odd numbers as

$2n + 1$ and $2m + 1$, where n and m could be any whole numbers.

Then students can argue that any time you add two odd numbers together, you will get the sum of

$$2n + 1 + 2m + 1 = 2n + 2m + 1 + 1 = 2n + 2m + 2 = 2(n + m + 1).$$

This means that the sum of two odd numbers is a multiple of 2, so it must be even.

MATHEMATICAL ARGUMENTATION IN MIDDLE SCHOOL

The statements start with what is known—a way to represent odd and even numbers—and the justification advances, in a couple of steps, to the truth of the conjecture.

These two example justifications point to how the process can lead to learning new concepts and skills. Say you are teaching sixth graders who are just learning to use algebraic expressions. Those who haven't fully assimilated this process will likely be more comfortable with something like dot diagrams for use in their justifications. If other students make the argument with variables, then you can elicit connections between the two that help all students better understand the representations. If students cannot make the symbolic argument themselves, then the lesson becomes an opportunity to introduce the use of algebraic expressions in this context, as well as an opportunity to introduce factoring of algebraic expressions. This is an example of how argumentation is not an add-on but a powerful way to engage students in content learning. We know that this is important to teachers, who have a great deal of content to cover between September and June. Argumentation shouldn't be one more thing to teach but instead should be part of a coherent practice that enables students to learn content and practices at the same time.

Chapter 4 further explains using variations on "How do we know it is true?" to begin justifications and provides ways to keep justifications going until they are complete.

Concluding—Closure on Truth or Falsity

The conclusion of an argument is agreement on the truth or falsity of a conjecture based on a justification. Concluding can be done by an individual, but in classroom argumentation, we are looking for the whole class to come to agreement based on the justifications that the students have made together or that are based on one student's argument that has been thoroughly critiqued by the other students. The process of concluding has two parts: (1) deciding whether a conjecture is true or false, based on its justification, and (2) summarizing the justification of the conjecture in logical order.

Leading a class in concluding can be challenging, so we provide several methods to accomplish it in Chapter 7.

ABOUT TRUTH

As teachers, it's worth thinking about what we mean by "true" in mathematics. There are at least two perspectives that can be taken. First, as we have done in this book, we can think of mathematical argumentation as a social process; we say that the truth is established by the agreement of a mathematics community—your class full of students. In this social-process view, mathematical truth depends on what people do together. On the other hand, mathematical statements are often thought of as true regardless of whether they have been accepted as such by any particular community. In this more absolutist view, that the sum of the measures of the interior angles of a triangle is 180° is not up for debate. But it is up for justification and shouldn't, in an argumentation-oriented mathematics class, be considered true until it is justified. For one thing, it depends on

Argumentation shouldn't be one more thing to teach but instead be part of a coherent practice that enables students to learn content and practices at the same time.

Concluding has two parts: (1) deciding whether a conjecture is true or false and (2) summarizing the justification in logical order.

how we agree to measure angles and define interior angles. Many of us tend to hold both views simultaneously—mathematics as enduring truth and mutually agreed-upon truth. You may not need to explain these points explicitly to students, but it's good to be aware of them.

What a class decides is true depends on what students already know. For example, a second-grade class acting as a mathematical community may conclude that "the sum of two numbers is always larger than either number" because students likely only know about whole numbers, for which the statement is true. But in middle school, they can revisit that statement once signed numbers have been introduced and refine their conclusion to refer to positive numbers only. At times, you may step in, representing the larger mathematical community outside the classroom. For example, students will probably need exposure to an introductory argument about the existence and meaning of irrational numbers before they can argue about such numbers for themselves.

TEACHING AS DISCIPLINED IMPROVISATION

It's clear that teaching for argumentation calls for thinking on your feet. Even with the example we've looked at so far, students' arguments for why the sum of two odd numbers must be even could come in quite a range, from pictorial to symbolic. You can't know for sure which will come up, and students may surprise you with an approach you hadn't considered. For this reason, we say that teaching for argumentation is improvisational. You may have seen improv performers on stage or television. The performers must work together to create a scene that has never been done before but that makes sense given audience cues. Teachers are certainly not mere performers, but they are called on to create something new each day—a lesson that must make sense and lead to learning while taking into account unforeseen student ideas that will be different from student to student, class to class, and topic to topic.

Disciplined improvisation relies on knowledge of mathematics, argumentation, and teaching moves to support argumentation.

Others have claimed, and we agree, that teaching calls for *disciplined improvisation* (Sawyer, 2004). We believe this term honors the complex work of teaching in which you engage. We've described what makes teaching improvisational. But improvisation does not mean anything goes—good improvisation in any field, from theater to music to teaching, builds on strong foundational knowledge and practices. Disciplined improvisation for teaching relies on several types of knowledge that are important: knowledge of mathematics, as seen through the perspective of teaching; knowledge of the structure and process of argumentation; and knowledge of teaching moves to support argumentation. This book addresses all these types of knowledge and will guide your own disciplined improvisational teaching of argumentation, yielding some predictability to the enterprise while helping you stay open to unforeseen student contributions.

Throughout this book, we provide opportunities for you to develop your disciplined, improvisational teaching by acquiring new teaching moves and building your own mathematics knowledge that is directly relevant to your teaching. Specific moves and categories of moves are presented in separate chapters for each part of argumentation. There are opportunities for you to try out the parts of argumentation for yourself with colleagues, as well as discussions of the mathematics relevant to the multiple classroom tasks you will find in each chapter.

IMPROVISATION FOR ARGUMENTATION AND NORM SETTING

Not only is teaching improvisational, but so, too, is mathematical argumentation. Cook (2015) describes the discovery of the famous mathematician and Fields medal winner Terence Tao:

> The ancient art of mathematics, Tao has discovered, does not reward speed so much as patience, cunning and, perhaps most surprising of all, the sort of gift for collaboration and improvisation that characterizes the best jazz musicians.

Both professional mathematicians and students must be prepared to build on the ideas of others in unforeseen ways as they uncover the mathematical truth together. As you well know, students may not come to your class with this view of mathematics. Supporting argumentation in your classroom requires developing new norms. A core set of norms needs to be in place for any argumentation to happen:

★ Treat each other with respect.
★ Speak so that everyone can hear.
★ It's OK to make mistakes; in fact, mistakes lead to learning.
★ Give your own ideas and build off the ideas of others.
★ Discuss ideas, not people.
★ We are finding out the mathematical truth together.

These norms are derived from work on accountable talk (Chapin, O'Connor, & Anderson, 2003). All the norms we suggest are indicated with stars (★) throughout the book.

> Whether it is a small group or whole class, I will remind disruptive students, "Let's check back on the norms: One person speaks at a time; be respectful."
> —Seventh-grade teacher

To help students develop a new view of mathematics as well as learn to engage in argumentation improvisationally, we provide specific norms for each stage of argumentation (Figure 1.4 shows a poster available for download at resources .corwin.com/mathargumentation) and promote the use of warm-up games that are based on the improv games that improv actors use to learn their craft. These games are widely used, and we have chosen and adapted them to address norms. There are many books, as listed in our references, and websites that you can find that describe improv games. Warm-up games have simple rules of interaction that foster spontaneity and strong collaboration among participants. They also encourage playfulness, which leads to the joy in argumentation. The games can be used at the beginning of selected class periods to introduce productive norms for that lesson. We have found that introducing the norms in the sometimes nonmathematical context of games can be important for students who may have never before participated in group discussions about the truth of mathematics. The games bridge between students' everyday experiences and an academic practice in a lively, engaging way that is safe even for those with little confidence in their mathematical competence. Each game should take 5 to 10 minutes at the beginning of a class period and will ultimately save you time if students experience a norm through a game. Teachers have found that a discussion to make explicit the connection between norm and game is essential. Playing the game grounds that discussion in students' own experiences, making it easier to establish a new way of behaving in mathematics class.

Brief *warm-up games*—even nonmathematical ones—are a fun, low-risk way for all students to become familiar with the norms of mathematical argumentation.

FIGURE 1.4 Argumentation Norms Poster

HOW TO DO MATH ARGUMENTATION

Generating Cases

- Think about more than just one case.
- Be creative:
  — Try simple numbers or shapes.
  — Try hard numbers or shapes.
  — Try "weird" numbers or shapes.

Conjecturing

- Use patterns to make statements about what will always be true.
- Make bold conjectures about what might be true.
- Avoid judging other people's conjectures.

Justifying

- Look for reasons why a conjecture is true or false.
- Build off of other people's ideas.
- Try to convince others of your ideas, but keep in mind that you could be wrong—which is OK.
- Show it a different way. Make a drawing, table, or graph.
- Be obvious.

Concluding

- Know when to stop.
- Retell the argument from beginning to end.
- Base your conclusions on what is said, not who said it.

 Available for download at
resources.corwin.com/mathargumentation

Zip, Zap, Zop is a classic simple warm-up game (e.g., Lobman & Lundquist, 2007) helps students understand that it is OK to make mistakes, speak so that everyone can hear, and pay close attention to one another.

ZIP, ZAP, ZOP

- Players stand in a circle so that everyone can see everyone else.
- The players throw an imaginary ball to one another within the circle, saying "zip," "zap," or "zop" (one with each throw, in that order, repeating the sequence until the game is over).
- The first player starts by throwing a "zip" to someone else in the circle.
- The catcher then becomes the thrower and throws the "zap" to someone else in the circle.
- Players continue, in any order, until most have had a turn.

Inevitably, someone will say the wrong word, and that can prompt what is known as the "circus bow." In the circus, when acrobats make a mistake and land in the net, they do not slink off in embarrassment. Instead, they jump off the net and make a bow with a flourish, as if they had intended the fall all along. Students can take a circus bow whenever they "mess up." This helps reinforce dramatically that making mistakes is OK.

Players in Zip, Zap, Zop must pay attention to each other to see where the "ball" is going for each turn. They need to speak loudly enough to be heard, but moreover, they need to look at each other. Looking at each other is so basic a norm that as adults we barely think about it, but students in mathematics class may be used to looking at the teacher only—except when they are getting into trouble! Looking at people as they speak helps establish an atmosphere of respect where everyone seeks to understand each other.

Here is the discussion one teacher had with her students after Zip, Zap, Zop.

Ms. Haddad:	All right. Think about what we just did. Think about how does the game relate to having conversations in the classroom? How does that game relate to us and our classroom norms? Who would like to share a connection for us? Jadzia?
Jadzia:	It connects because when we were doing Zip, Zap, Zop, we were going back and forth, and it's like speaking but actually is a game. And when we were doing the conjecture, argument, we are also doing the same with ideas instead of using zip, zop, or zap.
Ms. Haddad:	Outstanding. So passing around the ideas just like we passed around zip, zap, and zop. Other connections, Kai?
Kai:	We was speaking loudly enough so everyone can hear.
Ms. Haddad:	Yes! Speak loudly enough so when you share your conjecture, everyone can hear and respond. Other connections from the voices we haven't heard from? Tamiko?
Tamiko:	We didn't always have to get it right.
Ms. Haddad:	Absolutely. Can you share a little bit more about that?
Tamiko:	We could mess it up, say the words in the wrong order, and we would just start again.
Ms. Haddad:	Was it okay?
Class:	Yes!
Ms. Haddad:	Yes, so when we're making our conjectures, it's okay if we are not right the first time as long as you try. Right? Usher?

(Continued)

(*Continued*)

Usher:	Because like, if discussion takes time, one person's going at a time and nobody's talking over each other.
Ms. Haddad:	Perfect.

In addition to the norms stated earlier, there are norms that are specific to each part of argumentation (Figure 1.4), and we provide warm-up games for establishing them in the chapter for each part.

ACCESS FOR ALL

SHARING MATHEMATICAL AUTHORITY

Argumentation has the potential to help students develop as powerful users and creators of mathematics, with your support, in a classroom community. This requires explicit attention to developing this power, especially for students who have been historically marginalized in mathematics classrooms because of their race, level of English fluency, gender, or learning differences. If your students haven't done much argumentation, they are likely to believe that the teacher or the textbook is the only source of mathematical truth. By helping all students make bold conjectures and take mathematics into their own hands, you help students learn to take on mathematical authority for themselves as opposed to being passive receivers of content (Boaler & Greeno, 2000).

The social nature of argumentation is key. Students' authority is not developed in isolation; students need to talk to and listen to each other in developing their justifications (Esmonde, 2009). And you have the opportunity to validate all contributions to an argument as important, even if they don't follow the path to a justification that you expected. You also play an important role in facilitating students' clear communication and ensuring that students come away from argumentation with clear mathematical ideas.

Conjecturing, justifying, and concluding each provide opportunity to remind students that they are in charge, as a class, of deciding what is true and what is false. Of course, you may need to step in with your own mathematical authority if students persist in supporting a false conjecture or using limited reasoning. Even when you present students with a conclusion different from theirs, you can recognize the authority of the class as a community of mathematicians interacting with the even larger group of mathematicians—that you as the teacher represent—making explicit connections between their thinking and that of mathematicians (Stinson, Jett, & Williams, 2013).

This shift in mathematical authority can help students develop positive identities as mathematics learners. A good resource on equity-based practices to encourage positive mathematical identities is *The Impact of Identity in K–8 Mathematics: Rethinking Equity-Based Practices* by Aguirre, Mayfield-Ingram, and Martin (2013).

The norms introduced in this and subsequent chapters can contribute to students' developing mathematical authority, as students parlay their experiences with the games into the realm of mathematical argumentation.

All the techniques in this book are designed to help you broaden participation in high-level mathematics. The more concrete success you observe from a variety of students, the greater your expectations of all students will be. And just as we recommend students work together in a classroom community, we encourage you to take on this work with colleagues; sharing your successes and challenges with your colleagues can be helpful to all of you.

Throughout the book, we provide advice for how to provide support for culturally diverse students, students with learning differences, and students whose first language is not English. This is indicated with an *Access for All* icon.

A little more about these students for whom we provide extra support: For those developing their English language skills, rather than describing students as having limited English proficiency, we emphasize that these students come to school with many resources, including their home languages. For brevity, we will refer to them as English Language Learners, or ELLs. Also, while teachers' experiences with students with learning differences include students receiving special education services, as in a vignette in Chapter 5, we by no means address the range of particular learning disabilities you may encounter, on which special education experts can provide advice.

GETTING STARTED WITH ARGUMENTATION

Although not every lesson you teach will include all four parts of the argumentation model, almost any lesson can and should include students' justifications. Below, we provide three steps for an easy entry into teaching argumentation, which you can implement while reading the book.

Step 1: Start by adding some justification to your existing lessons.

> Use small questions—try to use them as often as possible and students will become aware that they can explain and justify. For instance, you can ask, "How do you know that's true?"
>
> —Eighth-grade teacher

No matter what lesson you teach, students are making mathematical statements—even if they are just the results of calculations. You can start having your students justify their statements simply by asking, "How do we know it's true?" Get students to give their best reasons based on the mathematical representations they have developed or that you provide for them.

For example, if you are teaching a lesson on division of fractions, students can use visual fraction models and stories to justify whether $\frac{3}{4}$ divided by $\frac{1}{2}$ is $\frac{3}{8}$ or $\frac{6}{4}$ (or something else). Whatever their answer to the calculation, you can ask, "How do we know it is true?" Students may start by explaining the procedure they remember—or misremember. You may need to prompt students in using representations to justify why the procedure works, asking, "Can you draw a picture that shows why?" or "Can you tell a story that explains why?" (For a complete framework on students' thinking about fractions, see Empson & Levi, 2011.)

In another example, a class is working on simplifying algebraic expressions. The teacher asks students why they could apply the distributive property to an expression: "How do we know that $3(x + 2) = 3x + 6$, no matter what x is?"

She provides time for students to draw pictures, reason with numbers instead of variables, and translate into real-world situations. The arguments are not very many steps long, but they provide insight into why the distributive property holds, instead of students having to accept it as just the way it's done in mathematics class.

Clearly, "How do we know it is true?" isn't the only question to ask to stimulate argumentation. You can go to Chapter 4 on justifying to see how to use different forms of "why" questions for slightly different purposes and to Chapter 5 to read more about the use of representations in justifications.

Step 2: Teach a lesson based on one conjecture.

You can begin a lesson with a "controversial" conjecture—a mathematical statement on which students may have differing initial views. If you know where you want the class to head, you can start with a conjecture that you have designed to highlight a new concept or insight. Some teachers have found it useful to start the class with a false conjecture, particularly one that represents an early conception that students commonly hold. Often, starting with a false conjecture is a good way to bring beginning arguers into the conversation. As you teach the lesson, students' statements about the conjecture will bring to light the reasons they hold their conceptualizations. As other students disagree and say why, the students can gradually reshape their ideas.

For example, you could start a lesson on multiplication of signed numbers with the following conjecture:

> Whenever you multiply any two negative numbers, the result is always negative.

You can elicit justifications from those who agree and from those who disagree, and you can contrast the justifications. You will have to make sure that students have some basis from which to justify. For example, you could ask them to use a number line to look at patterns formed by taking products of a sequence of numbers from 3 to −3, each times 2, and then have them look at the patterns that must hold for the same sequence, each multiplied by −2. This justification relies also on the fact that operations on signed numbers must be consistent with each other.

Alternatively, you could ask students to consider a justification using the distributive property and the fact that the sum of a number and its opposite is zero by asking, "What are two ways to evaluate the expression $-4(-5 + 5)$? How can they help us find the product of −4 and −5?" Even when offering these possible representations, you can press students to provide the justification themselves.

The justification is that the expression helps us this way:

> By doing what is in the parentheses first, we get 0. By using the distributive property, we get $-4(-5) + [-4(5)]$. If we already know that negative times positive equals -20, then by substituting we get $0 = -4(-5) - 20$. So $-4(-5)$ must be positive 20.

Chapter 3 offers more advice about starting a lesson with a single conjecture.

Step 3: Teach a lesson that requires students to create their own conjectures.

As you move into Step 3, you'll have students engaged in all four parts of the argumentation model. Keep in mind, though, that not every lesson will contain every part of the model. You may, for example, have students generate cases and make conjectures one day, and choose conjectures to justify the next. To begin a set of lessons based on an activity such as those provided in this book, you may want to entertain a lot of conjectures, which will come up for justification in different parts of subsequent lessons.

Chapters 2 and 3 provide classroom activities in which you elicit conjectures from students through generating cases. Use the lesson planning advice we give in Chapter 8 to help you prepare for a sequence of argumentation lessons.

ARGUMENTATION LESSONS VERSUS ARGUMENTATION *IN* LESSONS

You may be wondering if you need special lessons to teach argumentation or if you can integrate argumentation into lessons and activities you are already using. Our answer: Both! To fully engage students in the process of argumentation implied by our model, you will need lessons that are devoted to argumentation, and they should always address important content standards as well. We provide many tasks for such lessons in this book. But in your everyday teaching, you will find that you can include justification in many lessons just by asking students to convince each other of their ideas. The chapters on justifying provide you with teaching moves for doing so.

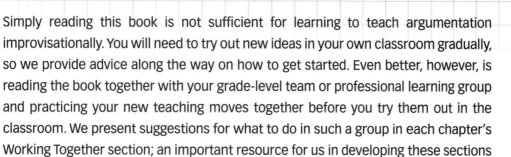

WORKING TOGETHER

Simply reading this book is not sufficient for learning to teach argumentation improvisationally. You will need to try out new ideas in your own classroom gradually, so we provide advice along the way on how to get started. Even better, however, is reading the book together with your grade-level team or professional learning group and practicing your new teaching moves together before you try them out in the classroom. We present suggestions for what to do in such a group in each chapter's Working Together section; an important resource for us in developing these sections was ATLAS (National School Reform Faculty, 2014).

(Continued)

(Continued)

Here's where to begin:

Exploration and discussion (30–45 min)

1. Convince yourselves: Why does the process for dividing fractions—the so-called invert and multiply—work?

   - Start with one example, say, $\frac{3}{4}$ divided by $\frac{1}{3}$ is $\frac{9}{4}$, and use a story or diagram to justify it.
   - Then using the example, discuss with your colleagues why the process always works.

Wrap-up and assignment (15 min)

2. Write down two or three questions to ask your students over the next few days. At the next meeting, report to the group about how the questioning went in your classroom. What worked? What was frustrating? (10 min)

3. During the week, commit to asking at least twice in one lesson, "How do we know it is true?" Briefly write down how you plan to respond to the justifications that students give and guide them toward making stronger justifications. (5 min)

NOTES

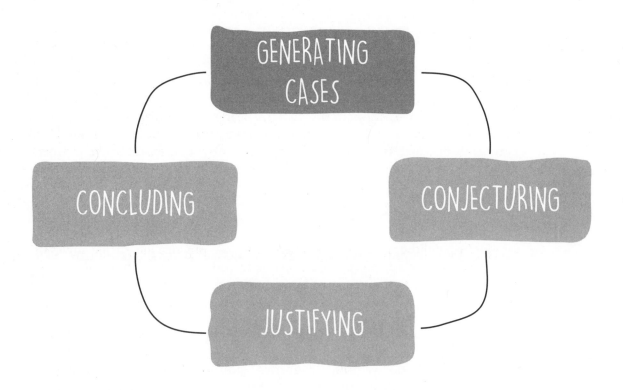

GENERATING CASES

CONCLUDING

CONJECTURING

JUSTIFYING

CHAPTER 2

Generating Cases

In this chapter, you will learn

- What generating cases means
- Through a vignette, how to support groups of students in generating cases
- Moves for setting the context, prompting for cases, and eliciting patterns
- Norms for generating cases along with games to help establish them
- How to plan for generating cases in your argumentation lessons
- Tasks for generating cases across different grade levels
- Through working with your colleagues, teaching moves that support the generation of cases

WHAT DOES IT MEAN TO GENERATE CASES?

When students generate cases, they create the mathematical examples—which could be numeric expressions or geometric shapes of a certain kind—from which they can both make conjectures and provide support for justifications. Cases are generated in order to see patterns. Patterns lead to generalizations that can be expressed as conjectures. In the justification stage, cases become important again as tools from which to reason. There will be more about cases in the chapters ahead, but in this chapter, we focus on the generation of cases that leads to conjecturing.

Generating cases can be a time for students to develop and practice mathematical skills. For example, students can practice finding sums efficiently as they look for patterns in those sums. For other tasks, they can practice measuring angles or, as in the following activity, using a coordinate grid to locate points. The tasks in this chapter provide opportunities for acquiring and building skills.

AN ACTIVITY RICH IN ARGUMENTATION AND CONTENT

Throughout the book, we present vignettes featuring Ms. Cooper and her sixth-grade class working their way through a three-lesson activity on argumentation in coordinate geometry. We provide the handout for that activity (see Task 2.1; download the handout at resources.corwin.com/mathargumentation) and for an optional GeoGebra activity to use with the activity. (Throughout the book, there are links to GeoGebra or Desmos activities that support some of the mathematical tasks in the book.)

With this activity, Ms. Cooper engaged students in all parts of the argumentation model and addressed content standards as well as other standards for mathematical practices—attending to precision, reasoning abstractly, and making use of structure. Her students were likely to have learned the conventions of graphing on the coordinate plane in fifth grade, but this activity provided the opportunity for her to review those conventions with her students:

- Two number lines that are perpendicular to each other are used as axes.
- The point where the axes intersect is called the origin.
- The horizontal and vertical axes are called the *x*- and *y*-axis, respectively.
- Each point on the plane can be named with a unique pair of coordinates (x, y), where the first coordinate is the distance from the origin along the *x*-axis and the second is the distance from the origin along the *y*-axis.

She helped students extend their prior use of the first quadrant of the graph, with positive coordinates, to all four quadrants when the axes are extended to include negative numbers.

She had these specific learning goals: Students will

- Graph polygons by plotting the coordinates of their vertices and connecting them

- Apply their knowledge of the coordinate system to explore the properties of lines that are parallel to the axes
- Apply and extend their use of numeric representations to developing algebraic representations that represent the overall pattern in the coordinates of all rectangles with sides parallel to the axes

These goals were closely tied to her state standards. Over the course of the three lessons, Ms. Cooper helped her students learn this important content *through* argumentation.

TASK 2.1 RECTANGLE COORDINATES

For this activity, use only rectangles with sides that are parallel to the axes.

1. On coordinate axes, make 4 rectangles of different sizes and locations. Label the vertices of the rectangles with their coordinate pairs.

2. What patterns do you see among the coordinates of the vertices?

3. Make conjectures about the relationships between the coordinates of the vertices of any rectangle whose sides are parallel to the *x*- and *y*-axes.

4. Justify one conjecture based on what you know about coordinate geometry. Remember, your conjecture may turn out to be true or false.

5. Write down your conclusion.

Use the GeoGebra activity for this task: https://goo.gl/DgEvXi. Created with GeoGebra (www.geogebra.org).

 Task 2.1 handout available for download at **resources.corwin.com/mathargumentation**

Coordinate geometry is a particularly important topic that merges two branches of mathematics: algebra and geometry. Through the juxtaposition of two intersecting number lines, and the identification of each point on the plane with a unique ordered pair of numbers, algebra can be used to represent geometric shapes from lines to curves and circles, and geometric shapes can also represent algebraic equations and functions. The power of this pairing supports the use of complex functions and shapes in later mathematics and all of STEM. Students' experiences with graphing on the coordinate plane, observing patterns in the numerical coordinates, and expressing those patterns algebraically are the first steps. Later in middle school, students will learn about the nature of the equations that represent lines, and they will also graph linear and nonlinear functions. In high school, they will learn about the equations for circles and the repeating curves of the trigonometric functions. In calculus, they will learn about further properties of shapes such as the area under a curve and how to express those properties symbolically.

VIGNETTE: SMALL GROUPS GENERATE CASES

The following vignette is based on the activity Rectangle Coordinates. Here is the generating-cases task from that activity:

For this activity, use only rectangles with sides that are parallel to the axes.

1. On coordinate axes, make 4 rectangles of different sizes and locations. Label the vertices of the rectangles with their coordinate pairs.

2. What patterns do you see among the coordinates of the vertices?

Before you read the vignette, take a few minutes to do the mathematics yourself. Using graph paper, draw axes and four rectangles, and label the coordinates of the vertices. Or you can use the GeoGebra activity we provide. Make sure you locate rectangles in all four quadrants. Think about how you chose the rectangles you did. Look for and note patterns in the coordinates.

In the following vignette, which represents part of the 45-minute period she allotted for this task, Ms. Cooper helped students choose cases and look for patterns among their cases through a whole-class introduction and then by visiting small groups. She made sure that, across the groups, students made typical rectangles and special ones such as squares.

Taking just a few minutes at the board with the whole class, Ms. Cooper established the use of the coordinate system with negative numbers, which were fairly new to her students, expanding their idea of a graph into four quadrants. She showed students how to plot a polygon by plotting its vertices and connecting them with line segments. But her instructional goals lay beyond this: She wanted students to understand *properties* of line segments and vertices of polygons graphed in the coordinate system, and she wanted students to eventually use variables to represent these properties.

Ms. Cooper then introduced the Rectangle Coordinates activity by posing a question to students.

Ms. Cooper: Now that we know about graphing with polygons with negative numbers coordinates as well as positive numbers, we can think about patterns. Do all the rectangles we can draw have some common properties? Let's start by just considering rectangles with their sides are parallel to the axes, just to start with some easy-to-think-about polygons.

Then she asked students to read Questions 1 and 2 on the handout.

Ms. Cooper: We are first going to "generate cases," and you'll work in your small groups. This means you are going to make four different rectangles, and you pay attention to the coordinates of each of their vertices—that means all of the corners. The rectangles and their coordinates will be your cases that will help you look for patterns. Try rectangles in different quadrants and try different dimensions for your rectangles. Be sure to record the coordinates of your rectangles in your notebooks. Then, look for patterns among the coordinates.

The students then started working in small groups. Ms. Cooper approached one group and asked what they were up to. She saw they had made two rectangles.

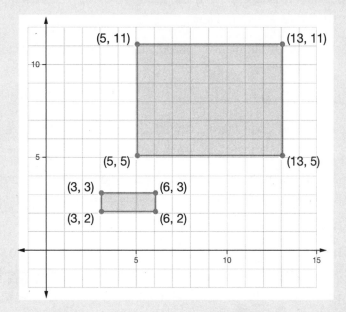

(Continued)

(*Continued*)

Sachi: We made a large rectangle and a small rectangle.

Ms. Cooper: OK, the instructions say to make four different rectangles. What other rectangles would be good to make?

Sachi: A medium-sized rectangle?

Ms. Cooper: Sure, you could try that. I notice that both of your rectangles have positive coordinates only. Try one that has negative coordinates as well.

Sachi, somewhat flustered, asked tentatively: Coordinates?

Ms. Cooper gently reminded Sachi and the group that they were making rectangles so that they could look at the patterns in the coordinates of the vertices. She pointed to the vertices and read one pair of coordinates to further direct students' attention.

Ying: Pattern? The large rectangle has large numbers and the small rectangle has small numbers.

Ms. Cooper: The large rectangle has large numbers. OK, which numbers are you referring to?

Ying: The side is large.

Ms. Cooper understood then that the group as a whole was a bit confused, and she clarified what she meant.

Ms. Cooper: What about the coordinates of the vertices of your rectangles?

She pointed to each of the vertices.

Sachi: The coordinates should be larger for the large rectangle because its sides are larger.

Sachi had stated a pattern, but Ms. Cooper didn't think it was based on Sachi's observations of coordinates of vertices. She told Sachi and the whole group,

Ms. Cooper: Try actually writing down the coordinates for the vertices of each rectangle, and see if that pattern holds. Don't forget to try a rectangle with some negative coordinates.

These rectangles would allow the group to probe Sachi's stated pattern.

Ms. Cooper approached a second group and saw that they had made squares.

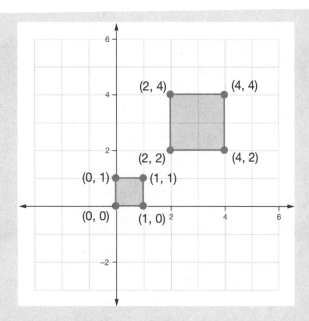

Ms. Cooper: Tell me about your cases.

Octavia: We made squares because we learned last year that squares are rectangles.

Ms. Cooper: So you are looking at the special cases of rectangles that are squares. That's great! You'll be able to look for patterns that may be special to squares—or may work for all rectangles. What else is special about these cases?

Queshaun: Well, I see a pattern of x- and y-coordinates being the same.

Ms. Cooper: Yes, each square has a vertex where the x- and y-coordinates are equal. How about if you try some squares where that is not true? Maybe try some over here. (She pointed to the third quadrant.)

The students started making a square with a vertex at (–3, –5), and Ms. Cooper moved on. She had challenged students to create rectangles that would help them see if the patterns they were noticing for their cases held generally.

At that point, Ms. Cooper decided to give the class as a whole some general advice on patterns and cases.

Ms. Cooper: As you generate rectangles and find the coordinates of their vertices, you're trying to find patterns in coordinates. Ask yourself, what is the same about these coordinates? What's different?

Aliyah: Should we try more rectangles and see if they work?

Ms. Cooper: Yes, you should do that—the instructions say to make four. Think about different types of rectangles to try. Be sure to try a few negative coordinates.

(Continued)

(Continued)

When Ms. Cooper visited another group, she noticed that they had made two rectangles so far and that each one had its center at the origin.

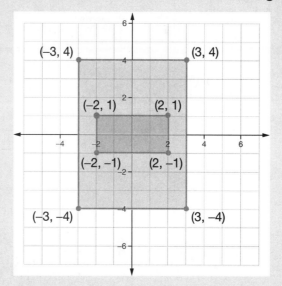

Ms. Cooper: Can you tell me about the cases you have chosen?

Malcolm: We tried one rectangle where the coordinates flipped.

Ms. Cooper: What do you mean by that?

Nadia: Well, the x-coordinate here is 3, and over here it is –3. So then we tried another one like that.

Ms. Cooper: That's a very interesting choice of cases. How do these two cases relate to the origin?

Lavar: They are balanced around the place where the axes cross.

Ms. Cooper then revoiced what Lavar had said in more formal mathematical language and encouraged them to continue to explore these special cases of rectangles.

Ms. Cooper: I see what you mean; we can say that the center of each rectangle is at the origin. Go ahead and make two more rectangles like that and you can report on the patterns you found for these special cases.

Ms. Cooper checked each group one more time to make sure that across the class there was a wide variety of rectangles being used and that every group had written down at least one pattern. Students had learned to use a coordinate grid that included negative numbers and had found patterns that could later lead to an exploration of properties of figures in the coordinate plane.

TEACHING MOVES

Ms. Cooper used a number of teaching moves that you can use yourself when you help students generate cases. You'll need moves for setting the mathematical context for the cases, eliciting more cases, and helping students see patterns among their cases.

Setting the Context

Setting the context means posing a question that establishes the situation that students will be exploring, discussing the task of generating cases with students, and making sure they understand what they should be doing. It's also a time when you can review or introduce mathematical conventions that students will need in generating their cases. You'll be able to reinforce these ideas as you check in with each group, so your initial instructions should provide just enough information to get students going on their cases.

After her students had read the task on their handouts, Ms. Cooper oriented them to what part of argumentation they were in by using the argumentation poster (Figure 2.1) displayed on the wall. She reiterated the written instructions so that any students who had trouble reading the instructions had another chance to comprehend them, and she asked if anyone needed clarification.

FIGURE 2.1 Mathematical Argumentation

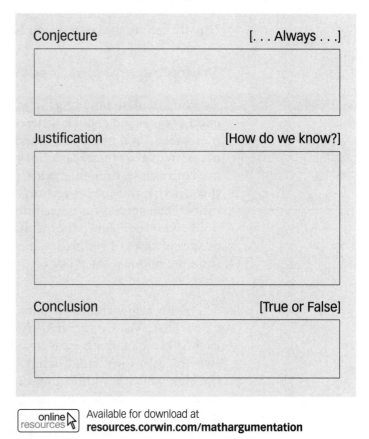

Available for download at
resources.corwin.com/mathargumentation

You can help students who may be struggling with mathematics skills while they are generating cases. Ms. Cooper took 5 minutes to review with students the conventions of graphing in the coordinate plane. Then, as she visited each group, she was able to check to ensure they could correctly read the coordinates of points. When one group showed some confusion about what coordinates are, she provided an example for them, providing just-in-time assistance to a group that was struggling. As you visit small groups that are generating cases, you'll be able to assess relevant skills and address them on the spot.

In summary, setting the context for generating cases involves

- Posing a question that engages students in mathematical thinking
- Ensuring that the class understands the mathematical requirements for the cases
- Reviewing and introducing conventions of mathematics needed for generating cases

Usually, setting the context will be brief—no more than 5 to 10 minutes.

Prompting for More Cases

As you circulate among groups, you'll want to make sure that students try enough cases and enough different types of cases. Students often choose examples based on numbers or situations with which they are most familiar. For many students, typical cases for rectangles are those with unequal sides. Across the entire class, though, students should examine a range of typical cases. For this activity, this means rectangles in different quadrants and spanning quadrants. Then special cases are important for both seeing if general patterns still hold and recognizing patterns that are specific to them. Squares are an example of special cases of rectangles, having four equal sides. We have more advice for thinking about what types of cases to prompt in the planning section of this chapter.

There are a few things to note about how Ms. Cooper prompted for more cases. First, Ms. Cooper had students work in groups so that they could help each other in finding patterns. As Ms. Cooper visited each group, she had two things on her mind: what kinds of cases the group could contribute to the class and what kinds of rectangles would challenge that particular group of students, in both using negative numbers (which were relatively new for her students) and finding patterns. Not all groups needed to make all types of cases, because Ms. Cooper had ensured that a wide variety would be made across the class.

Two of Ms. Cooper's groups were generating special cases of rectangles, and she encouraged them to continue to make more of these special cases. Squares with vertices at points (a, a) and (b, b) can be made with only two numbers across their coordinates. Rectangles that are centered at the origin enable the insight that reflection over the y-axis changes the sign of the x-coordinate, and likewise for the x-axis, the signs of the y-coordinates change. Ms. Cooper wanted to take advantage of students seeing patterns in these special cases of rectangles as material for the whole class when making conjectures.

Moves for prompting for more cases include

- Having students work in small groups, which reduces the number of cases any individual has to generate
- Helping students think of special and typical cases to examine
- Seeding a variety of cases across groups
- Seeding more and less challenging cases with different groups

Eliciting Patterns

Finding patterns is the transition between generating cases and conjecturing. Statements of patterns may be fragmentary or use students' own notation. That's OK. Students will make more complete statements as conjectures, as you will see in the next chapter. Statements of patterns describe commonalities among the cases, but students may not generalize to all rectangles; that comes with conjecturing.

Students may need help in finding and expressing patterns across their cases. In the vignette, Ms. Cooper gave the whole class some general advice on finding patterns by asking what was the same and different for their cases. However, one group seemed to be looking for patterns right away before they had finished making the required number of cases. This is good, because the patterns they found along the way could suggest new cases to try—those that would confirm or break the pattern. If students have trouble finding or articulating patterns, there are specific moves you can try.

To elicit patterns, you can ask students,

- "What do the cases have in common?"
- "What is the same about every case?"
- "What changes from case to case?"
- "What patterns can you find when you look at *all* the cases?"
- "Are there any cases that break the pattern you think you may have found?"
- "Do the cases fall into categories?"

ESTABLISHING NORMS

In this section, we provide norms for generating cases and warm-up games that will help you establish those norms with your students.

Norms

Generating cases has some commonalities with brainstorming. These are some norms that can help students understand what is expected of them during this phase of argumentation:

- ★ **Make multiple cases.** In order to see patterns that will lead to conjecturing, it is often useful to try more than one case. This is particularly important for students who consider only special cases—such as only considering squares when arguing about quadrilaterals.
- ★ **Be creative: Try simple numbers or shapes. Try difficult numbers or shapes. Try "weird" numbers or shapes.** Simply telling students to be creative may not be sufficient for them to know what to do. This norm includes specific ideas about what being creative means when generating cases for argumentation.

Warm-Up Games

We include two games designed to establish these norms so that your classroom has an atmosphere of generativity and creativity. The games are not necessarily mathematical and shouldn't be thought of as frivolous; instead, they should be done as a way for students to warm up their creative thinking. Each game should take 5 to 10 minutes of class time, and the time spent will increase your efficiency in establishing norms. The games also help establish the more basic norms, such as "mistakes are OK" and that mistakes actually enhance learning. It's important that after playing a game, you have a brief discussion with students to explicitly connect the game with the norms.

FIRST LETTER, LAST LETTER

Players stand in a circle so that everyone can see everyone else.

©iStockphoto.com/kali9

- The teacher calls out a category, such as "animals."
- The first player throws an imaginary ball to another player, at the same time saying a word in that category: for example, "tiger."
- The person who receives the ball repeats the word "tiger" and then throws the ball to another player with a new word, starting with the last letter of the first word: in our example, "rabbit," for "r."
- The third person catches the word "rabbit" and then sends a new word starting with the letter "t": for example, "toad."
- The game continues until all players have had a chance to catch a word and say a new word.
- If a player repeats a word that has been said, the group can decide if that deserves a circus bow or whether repeating words is allowed.

Variations: For large classes, make two or more smaller groups and follow the same rules.

In First Letter, Last Letter (e.g., Hall, 2014), students must generate multiple responses under the same set of rules. This is analogous to how students generate cases that follow certain rules or meet certain constraints—for example, in the vignette, not just any coordinates but also the coordinates of rectangles parallel to the axes. Students have to give themselves freedom to come up with new cases while staying within the constraints, and they should recognize that there are always more cases to be made. In the game, the given category and last letter are the constraints, and the words the students say are the cases.

The following game helps students use their imaginations while collaborating. Students should have the opportunity to approach mathematics as a creative endeavor, as mathematicians do. A creative mindset will be very useful when generating cases and making conjectures about abstract ideas.

OFF THE SHELF

- Players stand facing a partner.
- Players take turns reaching up to an imaginary shelf.
- Without thinking too much, players say what imaginary item they pulled off the shelf.

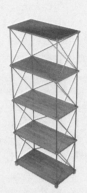
©Clipart.com

The pretend nature of Off the Shelf (e.g., Johnstone, 2012) allows students to practice engaging with ideas, making an effort to collectively "see" what's not there. This is really useful when doing any of the mathematical argumentation stages because it helps students see that ideas can be put forth and become objects we discuss and with which we interact.

PLANNING

In preparing for generating cases, you'll need to choose appropriate tasks and consider how many cases of what types you'll want students to make.

Selecting Tasks

If you are planning your own argumentation activity, choose initial tasks in which students can examine a variety of cases and look for patterns. The tasks presented in the next section are good examples of generating cases that are good for argumentation. You can find similar tasks in your textbook by looking for the word *pattern* to cue you in and also by looking for general statements about shapes, numbers, or situations with which students have some basic familiarity or some procedural fluency. For example, one teacher had students generate cases to investigate whether a fraction raised to a power always resulted in a smaller fraction. This worked because students knew how to multiply simple fractions, for example: $\frac{1}{2}$ times $\frac{1}{2}$ times $\frac{1}{2}$. Starting with the familiar, argumentation is a way to help students develop new concepts and methods, for example, as students represent numeric patterns algebraically, as Ms. Cooper's students will do.

Generating cases based on operations on rational numbers or familiar geometric shapes works well for middle schoolers, though it may be helpful to stick to whole numbers or integers for some tasks. Choosing a set of algebraic equations across which to generalize would likely *not* be a good choice for students who are just beginning to become fluent with variables and expressions. Instead, you can help students build their understanding of algebra by generalizing from a set of numeric cases.

For example, Ms. Cooper began with the first question in the Rectangle Coordinates activity, which asks students to generate cases of rectangles parallel to the axes. These rectangles make for good cases because the coordinates are fairly straightforward to determine. The cases then serve as vehicles for students' insights that take them beyond their initial understanding to developing new or deeper understandings, as you will see in the transition from numbers to algebra found in the following chapters.

ACCESS FOR ALL

You can select tasks that help students make the transition from arithmetic into algebra, where they use variables to represent patterns. Do not assume, however, that students must be able to rapidly perform operations on numbers before they can do algebra. Procedural fluency means understanding the meaning of as well as when and how to use procedures, which can include using a calculator instead of paper-and-pencil algorithms (National Council of Teachers of Mathematics, 2014; National Research Council, 2001). This kind of procedural fluency makes it easier to see the numeric patterns that can be represented with algebraic symbols.

Sometimes it makes sense to provide students with cases, for example, a variety of triangle cutouts for a lesson on the sum of the measures of the angles in any triangle. Giving students premade cases is one way to focus students' attention on the work of seeing patterns rather than the potentially lengthy task of creating the cases. This must be balanced with the fact that having students create their own cases provides them with the opportunity to make decisions about the features of the cases they need in order to see patterns.

How Many Cases of What Types?

You'll need to think about the different types of typical cases that you will want students to explore. There have to be enough types of cases to provide some grounds for generalizing (Williams et al., 2011). In the Rectangle Coordinates activity, for example, two rectangles would really be a sufficient number of typical cases since coordinates work consistently everywhere in the plane. Students won't know this, however, so they need to look at different categories of rectangles to be convinced that the patterns hold, for example, rectangles with a variety of side lengths, in all four quadrants, and spanning quadrants.

You should also consider special cases that you will want students to examine. In the Rectangle Coordinates activity, special cases include rectangles where only three numbers are required; these are rectangles with two different side lengths and with one vertex for which the x- and y-coordinates are the same. As we saw in the vignette, squares that also have a vertex for which the x- and y-coordinates

Do not assume that students must be able to rapidly perform operations on numbers before they use variables.

are the same require only two numbers to make their coordinates. Other special cases can yield their own patterns, as Ms. Cooper and her students saw with rectangles centered at the origin.

What makes for a typical or special case can be subtle. Mathematicians use this reasoning process in their work, and you can help students develop this mathematical practice. For geometric tasks, a regular polygon can be a special case. In numeric contexts, 0 and 1 are often special cases because they are the identity elements for addition and multiplication. Special cases can be useful for testing extremes—"Does the pattern hold even in this case?"—but generalizing only from special cases can be problematic. For example, 2 also is often a special case because the fact that $2 \cdot 2 = 2 + 2$ can give rise to patterns that do not hold for other numbers.

ACCESS FOR ALL

You can differentiate instruction by choosing more or less challenging cases for students to work with. But also consider the core mathematics to which you want every student to have access. For example, in the Rectangle Coordinates activity, working with coordinates requires dealing with signed numbers. You can ask students who are working at a more basic level to concentrate on rectangles in a single quadrant, and ask more advanced groups to consider cases that span quadrants, where determining the signs of the coordinates is more challenging. To ensure that all students get a chance to work with some negative and positive numbers, make sure every group works on at least one rectangle in the second or fourth quadrant.

Helping Students Record Their Work

The generating cases phase is a good time to start helping students record their work in writing, in an organized way in their mathematics notebooks, taking ownership of their work. Ms. Cooper's students could record the coordinates in a table as well as see them on the graph, which would help reveal patterns. We have more to say about representations in Chapter 5, but suffice it to say here that different representations—tables, graphs, diagrams—can help students recognize patterns in their cases.

In summary, when planning for generating cases, ask yourself,

- What patterns are likely to be seen in this activity?
- How many cases of what types will students need in order to see and be confident in patterns across the cases?
- What are the typical cases for this task?
- What are the special cases that are important to consider, including extreme cases?
- Which cases can students who are working at different levels be expected to generate?
- What teaching moves can I use to help students generate a good set of cases?
- Which representations will help students organize their cases?

TASKS FOR DIFFERENT GRADE LEVELS

What follow are some good tasks for generating cases. The handouts on the website contain support for all four parts of argumentation.

TASK 2.2 FOR STUDENTS LEARNING OPERATIONS ON SIGNED NUMBERS

> Try finding sums of positive and negative numbers and notice whether the sum is positive or negative. What patterns do you find?

 Task 2.2 handout available for download at **resources.corwin.com/mathargumentation**

For this task, have students use a number line or double-sided counters to work out the sums. Be sure to have students try sums where the absolute value of the negative number is greater than that of the positive number, as well as the reverse (e.g., –4 + 3 and 4 + [–3]). There is one basic pattern for students to find:

> The sign of the sum of two numbers is the same as the sign of the addend with the greatest absolute value.

A special case occurs when the absolute values of the addends are the same; then the sum is 0.

TASK 2.3 FOR STUDENTS LEARNING ABOUT MAKING TRIANGLES FROM GIVEN CONDITIONS

> Create line segments of various lengths. Find sets of three segments that form triangles. What patterns do you see that relate the lengths of three segments that can be used to form a triangle?

 Task 2.3 handout available for download at **resources.corwin.com/mathargumentation**

This task will provide students an opportunity to state patterns among triangles in terms of their side lengths:

> When the sum of the lengths of the two shorter sides is less than the length of the longest side, then you can't make a triangle.

> When the sum of the lengths of the two shorter sides is equal to the length of the longest side, then you can't make a triangle.

> When the sum of the lengths of the two shorter sides is greater than the length of the longest side, then you *can* make a triangle.

Students may not see three types of cases right away. You can draw their attention to them. Often there is a critical case between two types of cases—for this task, the case where the sum of the shorter sides is equal to the longest side (Figure 2.2b) distinguishes the cases where the sum is greater (Figure 2.2c) or less (Figure 2.2a) than the longest side. For some numeric tasks, zero is the critical case between positive and negative.

Use the GeoGebra activity for this task: https://goo.gl/YIuz4K. Created with GeoGebra (www.geogebra.org).

FIGURE 2.2 Three Types of Cases for Lengths of Segments
That Do Not and Do Form a Triangle

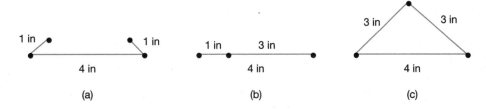

(a)　　　　　(b)　　　　　(c)

Students can directly manipulate line segments in which they can compare lengths without measuring. In the GeoGebra activity, students drag vertices or use sliders to change lengths of sides; the measurements of the sides are provided. We have offered students uncooked spaghetti strands that they can break into three lengths and test to see if they make a triangle, which avoids the need for measurement. If you want students to use measurements, you can provide rulers or a handout with lines drawn and labeled with lengths, as on the activity handout.

TASK 2.4 FOR STUDENTS LEARNING TRANSFORMATIONS IN GEOMETRY

Plot the vertices of a triangle. Reflect it over the *x*-axis. What patterns do you see in the coordinates of the vertices of the triangle and its image?

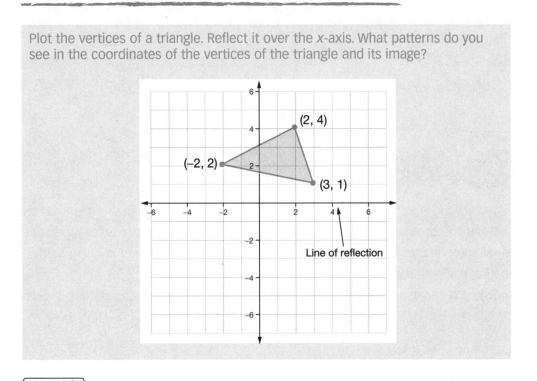

Use the GeoGebra activity for this task: https://goo.gl/ruy3v0. Created with GeoGebra (www.geogebra.org).

online resources ⬚ Task 2.4 handout available for download at **resources.corwin.com/mathargumentation**

You can have students find the reflected image using informal ways that reflection can be introduced, or students can use our GeoGebra activity, which will show the results of a reflection. This activity will lead to statements of the pattern that

> Only the *y*-coordinates change signs when triangles are reflected over the *x*-axis.

Students may be ready to state the pattern algebraically about each point in the plane:

The image of (*a, b*) reflected over the *x*-axis is (*a, –b*).

You can, of course, also have students reflect shapes over the *y*-axis and look for patterns.

DIGITAL TOOLS

For Rectangle Coordinates, you can use our GeoGebra activity or you can create your own with dynamic geometry software that you already use. Your page or file should have a premade rectangle with coordinates of vertices displayed, which can be reshaped and repositioned by dragging the vertices. For some students, including those with identified learning disabilities, plotting many points and keeping track of the order of *x*- and *y*-coordinates can be a barrier to participation in the activity (Sweller, 1994). Yet, there is no reason that they can't look for patterns and then participate in argumentation, even without performing the mechanics of plotting and reading off points. For all students, being able to observe changes when a point is dragged provides access to many more cases than they could plot by hand.

 WORKING TOGETHER

As you work together with other teachers in a professional learning community, department, or grade-level group, you can develop and hone your teaching moves for supporting the generating of cases. These activities engage you in doing mathematics, planning based on the mathematics you uncover, coming up with teaching moves you can use for generating cases, and, finally, trying out your moves in your classroom.

Exploration and discussion (45 min)

1. As a grade-level group, choose a task or use one of the following tasks, and do it together to generate cases and find patterns.
    a. For sixth grade: Rectangle Coordinates on page 21
    b. For seventh grade: the signed numbers task on page 34
    c. For eighth grade: the transformational geometry task on page 35 (15 min)

2. For your chosen task, answer three questions for planning a lesson:
   a. What patterns might students see?
   b. About how many cases should students examine, and across what categories?
   c. What are special cases you want to have students consider or avoid? (15 min)

3. Review the teaching moves presented in the chapter. Come up with your own moves—questions or prompts—to use in your lesson for the following purposes:
   a. To get students to generate additional cases
   b. To encourage different groups to generate cases at their skill levels
   c. To push students to consider and generate special or extreme cases (15 min)

Wrap-up and assignment (15 min)

Select a task. Then outline a plan for having students generate cases. For example,

- Have students generate cases in pairs or small groups.
- Next, have each group make a list of a few patterns.
- Finally, post them on a wall for the whole class to see.

Note: In order to complete argumentation lessons on these tasks, you will need to plan for conjecturing, justifying, and concluding as well. You will learn more about the process in subsequent chapters.

NOTES

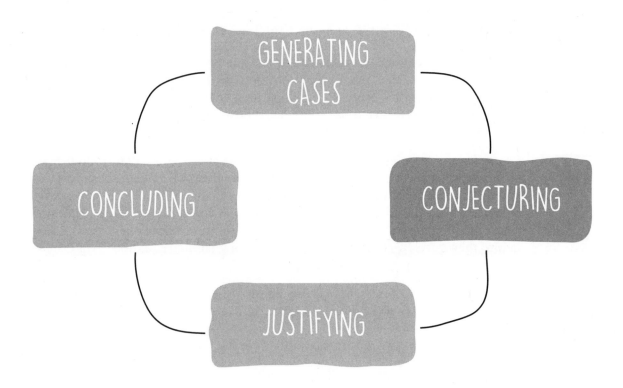

CHAPTER 3

Conjecturing

In this chapter, you will learn

- What conjecturing means in the mathematics classroom
- Through a vignette, how students create conjectures
- Moves for explaining, clarifying, and helping students extend each other's conjectures
- Moves for supporting conjecturing in the "if . . . then . . ." form
- Norms for conjecturing along with games to help establish them
- How to plan for conjecturing in your lessons
- Tasks for encouraging conjecturing that can be adapted for different grade levels
- In working with your grade-level team or department, how different types of conjectures can engage your students in argumentation

WHAT DOES IT MEAN TO CONJECTURE?

As Ms. Cooper tells her students in the following vignette, a conjecture is a mathematical statement that you think might be true, based on what you know so far. Conjecturing is the process of making these statements without yet deciding if they are true. In this chapter, you'll learn how to guide students through that process.

Conjectures can be general or specific. General conjectures often follow from generating cases and noticing patterns—when students state what those patterns are, they are on their way to making conjectures that generalize from a few cases to what is true for all cases. For some activities, conjectures will be more specific. For example, students might conjecture a solution to one example equation before they develop general methods for solving similar equations. We discuss both kinds of conjectures in this chapter.

A good conjecture should contain precise language with the purpose of making the conjecture clear to everyone—but it's unlikely that students' first attempts will come out this way. That's why clarifying conjectures is important. We explain more about clarifying conjectures in this chapter.

As important as the process of conjecturing is, there are times when you will want to start out with a premade conjecture. We discuss that approach and when to take it.

VIGNETTE: CONJECTURING TOGETHER

The following vignette is based on the conjecturing task from the Rectangle Coordinates activity.

> Make conjectures about the relationships between the coordinates of the vertices of any rectangle whose sides are parallel to the *x*- and *y*-axes.

Before reading the vignette, take some time to make conjectures about what you think is always true about coordinates of vertices of all rectangles with sides parallel to the axes or what might be true just for the category of special cases you made. Consider also false conjectures that students might make.

Ms. Cooper structured a whole-class discussion leading to a broad variety of conjectures based on the patterns students had found. Through this process, she addressed important content: Students were exploring the properties of lines that are parallel to the axes through making conjectures about the vertices of sides of rectangles. She was preparing students to use algebra to describe coordinates in subsequent lessons by making general conjectures in words.

After every group in the class had made four rectangles and found patterns in the coordinates, Ms. Cooper prompted for conjectures about those coordinates.

Ms. Cooper: Based on the patterns you found, what conjectures can you make about the coordinates of vertices of rectangles?

Yates: What do you mean by conjecture, again?

Ms. Cooper: A conjecture is your best guess about what is true, mathematically. It's a mathematical statement that you make, that you think might be true, based on what you know so far. For your conjectures today, what you know so far are the patterns you saw in the coordinates of the vertices of the rectangles you made.

Then she restated her question:

Ms. Cooper: Based on the rectangles you have made and thought about, what is something you can say might be true, or seems to be true, about the coordinates of the vertices of all rectangles?

Several students raised their hands, and Ms. Cooper called on Ying.

Ying: The numbers get bigger or smaller for different rectangles.

Ms. Cooper wrote this on the board and kept asking for conjectures. She got these two:

1. The numbers get bigger or smaller for different rectangles.

2. There are four different numbers used but in all different orders.

The class seemed to be out of conjectures at that point; no one raised a hand. Ms. Cooper was ready to prompt for more.

Ms. Cooper: OK, there are four different numbers. Did you see any patterns there? For example, what stayed the same?

Malcolm: We saw that for the up-and-down sides, there were two coordinates that matched.

Ms. Cooper added that conjecture to the board:

3. For the up-and-down sides of the rectangles, there were two coordinates the same.

Ms. Cooper: They just made a conjecture about one pair of sides in rectangles. Can anyone make one about the other pair of sides?

(*Continued*)

Several hands went up, and the class got a new conjecture for the board:

4. For the horizontal sides, there are two *x*-coordinates that are the same.

Ms. Cooper noticed that this conjecture was false, and she wanted to make sure it was recorded, because a false conjecture would turn out to be useful during justifying. But she didn't announce this so that students could justify it for themselves. After this, the class seemed more confident, and several more conjectures came up pretty quickly:

5. The opposite corners don't have any numbers in common.

6. Large rectangles have coordinates that are farther apart than small rectangles.

7. All four vertices have two numbers each.

Conjecture 7 at first seemed very obvious to Ms. Cooper, for students who had been graphing for at least 2 years, but she accepted it as a contribution to the list, especially because it came from a student who was usually quiet. Then she prompted the class again, this time for patterns based on special cases.

Ms. Cooper: What about those of you who looked at some of what we called special cases. Did you see any patterns that worked for them?

Three more conjectures went up on the board:

8. There are just two different numbers used to make the coordinates, sometimes.

9. When a rectangle crosses the *y*-axis, it has a positive *x*-coordinate and a negative *x*-coordinate.

10. To make long skinny rectangles, two numbers in the same pairs of coordinates should be close to each other and two numbers in the other pair of coordinates should be far away from each other.

11. When rectangles are centered around the origin, the signs of the coordinates flip.

Even though Ms. Cooper asked for "your conjecture," she didn't put any names on the board next to conjectures. When she was done, the board contained 11 conjectures, which she had written as students said them, without much commentary. She noted that the conjectures were closely related to the patterns students had recognized, although some nicely built on others' conjectures. She ensured that at least one student from each group spoke. To do that, she called on some students who hadn't raised their hands, asking,

"Can you state a pattern that you found as a conjecture?" and "Does this conjecture help you think of what else might be true?"

Almost all of the students' conjectures needed some work to be mathematically precise, that is, to say exactly what mathematical situation they were describing.

Ms. Cooper: Great work making so many conjectures! Now we are going to clarify the conjectures. We want to make sure that everyone understands exactly what they mean, and to do that, it will help us to use mathematical language precisely.

She read Conjectures 3 and 4 aloud and asked if anyone had any questions about them, in order to clarify them.

Aliyah: Is horizontal a mathematics word? I can never remember if that is up and down or across.

Bailey: Horizontal is across. I think we should change up-and-down to say vertical, and then the two conjectures match.

Ms. Cooper replaced "up-and-down" with "vertical," and at the side of the display board, she sketched a pair of axes and wrote the words *horizontal* and *vertical* correctly on each axis. She also labeled them "*x*-axis" and "*y*-axis." Then all students could refer to the sketch, as needed. She asked for more clarifying questions.

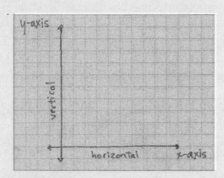

Camden: Shouldn't Conjecture 2, that one about four numbers, say something about coordinates? Or rectangles? That's what we are making conjectures about.

Ms. Cooper: Aliyah, could you restate the conjecture using additional mathematical language in order to clarify it?

Aliyah: The coordinates have four different numbers, in different combinations.

Ms. Cooper: Could we even say which coordinates we are talking about? Remember, we are working with rectangles.

Dale: The coordinates of the vertices of the rectangles, like you said.

(Continued)

> Ms. Cooper modified the conjecture and asked if anyone wanted further clarification. They ended up with
>
> When you write down the coordinates of the vertices of any rectangle (parallel to the axes), you will use four different numbers, but in different orders for the different coordinate pairs.
>
> Gabriel: When you put it that way, I don't agree with that conjecture, because I found a rectangle with only three different numbers in the coordinates.
>
> Ms. Cooper: We are only clarifying conjectures at this stage, not deciding if they are true or false. That comes during justifying, which we will do next. That's when we'll consider Gabriel's rectangle, as part of the justification of whether this conjecture is true or false. Let's continue with clarifying for now.
>
> Finley: For Conjecture 5, we could change the word *corners* to be more "math-y."
>
> Ms. Cooper asked Finley to go ahead and restate the conjecture, and she revised it on the board, using the word *vertices* instead of *corners*.
>
> The opposite vertices don't have any numbers in common.
>
> Octavia raised her hand eagerly, and Ms. Cooper called on her.
>
> Octavia: Well, I say, pick any two points, then connect them. As long as that line is *not* parallel to the axes, then I can always make a rectangle with them as the corners. That's a conjecture, isn't it?
>
> Ms. Cooper: It's a statement about what is true for *any* two points and rectangles, so yes it is!
>
> Ms. Cooper wrote this 12th conjecture on the list.

Conjecturing, including clarifying, had taken about half a class period. Ms. Cooper thought it was time well spent, because the students expressed mathematical ideas and were working on precise language—and all her students had been engaged. In terms of progressing through content, students had more than met the standard on graphing line segments and polygons. She also recognized that with some additional questioning in a subsequent lesson, she could build on Conjecture 11 to address a standard about reflections of points over the axes.

TEACHING MOVES

Ms. Cooper used a number of teaching moves that you can use yourself when you help students make conjectures, and we discuss some additional categories of moves as well. You'll need moves for broadening access to and explaining conjecturing; eliciting, clarifying, and building on conjectures; and supporting students' use of precise mathematical language in their conjectures.

Broadening Access Through Conjecturing

Ms. Cooper provided access to mathematical authority with several techniques specific to conjecturing. Knowing what cases students had generated in small groups, she was able to call on a student from each group to make a unique contribution to conjecturing. She scaffolded additional conjecturing so that more students could bring in new ideas when classroom conversation seemed stalled. She drew attention to the content of conjectures rather than who made them. She accepted initial conjectures in students' own language, knowing she could help make them more precise in the clarifying stage. Finally, she affirmed a spontaneously made conjecture that didn't fit with her lesson plan.

Explaining Conjecturing

There's no reason that students would come to your class knowing the term *conjecture*. And there's no reason to keep it a mystery. Teachers with whom we have worked have found a number of ways to introduce students to the term and to the practice of conjecturing.

Ms. Cooper began with three sentences explaining what conjecturing is, how it is done, and what it would look like in the activity.

Teachers have also made connections between conjecturing and predictions students may have made in their own lives, for example, when they predict which team will win a big game. That example makes clear that it's important to reason from what you already know to what might be true. The point must also be made that a conjecture is different from a prediction about the game because while statements about mathematics are enduringly true, the outcome of the game is not and depends on many factors in the moment.

Another teacher had students make initial observations about solutions to linear equations, such as, "for $\frac{1}{2}x + 3 = 5$, x must be a number greater than 2," and *then* she labeled these statements as conjectures. She didn't tell students in advance what a conjecture was but rather labeled something students had already done as conjecturing. Any of these introductions can work.

Eliciting Conjectures

You can encourage students to make a wide variety of conjectures, including bold conjectures that students have only a hunch are true. There are three reasons for doing so: First, it's good for students to stretch their mathematical imaginations. This can be a motivating factor for many students, particularly those who have been turned off by a perception of mathematics as a set of procedures to be memorized. Second, authentic mathematical argumentation takes place when the truth of a conjecture at hand is really unknown. Third, argumentation provides opportunities for students to learn concepts, and the broader the set of conjectures a class makes, the wider the net cast for concepts to be learned.

How do you encourage many and bold conjectures? The simplest prompt is in a shift from what *is* true to "What do you think *might* be true?" But there is more to be done than that. This is a time to ensure broad participation from the class. The more voices that are heard, the wider the range of conjectures is likely to be. So any techniques you already use for including more students in the mathematical conversation—such as "think-pair-share"—are likely to be useful in the conjecturing phase. Individual thinking time can be important, so students can carefully study the cases that have been generated. You can invoke norms for argumentation: You can explicitly ask students to build on the ideas of others by having them look at other conjectures and asking if that helps them think of a new conjecture. The process of conjecturing as a group can lead to more conjectures as students hear each other's words. This requires that students really listen to each other—another of the basic norms for argumentation from Chapter 1.

For example, in the vignette, a student made a bold conjecture (in more precise language here):

> Pick any two points so that the line segment connecting the two is *not* parallel to the *x*- or *y*-axis. I can always make a rectangle parallel to the axes with vertices at those points.

This conjecture went beyond the pattern recognition students had done in generating cases and reflected an insight that may have come from listening to others' conjectures.

When conjecturing is particularly challenging for students, or if students are new to conjecturing, you can, as Ms. Cooper did, gather the whole class together to conjecture based on the cases that small groups generated. This makes it easier to introduce specific mathematical concepts or representations as tools for conjecturing. Ms. Cooper had another reason for making conjecturing a whole-class activity. She wanted the conjectures to belong to the entire class instead of being identified with individuals, because it was early in the school year and her students were still reluctant to make a conjecture that might turn out to be false. While she did make sure that conjectures came from each group, she didn't make a big deal out of it publicly.

There are other strategies you can use when students have trouble moving from conjecturing about what *is* true to what *might* be true. For example, you can ask them for conjectures in categories:

- "What are you absolutely sure of?"
- "What are you pretty sure is true?"
- "What could possibly be true?"
- "How about what is false—can you make a conjecture that you are pretty sure is false?"

The last can be a confidence builder for students who may be unsure of what to conjecture and sets students up for simple justifications. You can even start the class off with a false conjecture yourself; just to set a "good" example.

Supporting Specific and General Conjectures

Whether students make specific or general conjectures depends in part on how familiar they are with a particular topic. Specific conjectures are appropriate when students are just learning, for example, how to add numbers (for elementary students) or how to use algebraic symbols (for middle school students). Figure 3.1 shows related specific and general conjectures.

FIGURE 3.1 Specific and General Conjectures

SPECIFIC	GENERAL
$3 + 5 = 8$	The sum of any two odd whole numbers is even.
For $x + 13 = 27$, $x = 14$	Any time you have an equation, you can subtract the same number from each side of the equal sign, and you get an equivalent equation.
The sum of interior angles for the triangle with 30°-60°-90° is 180°.	For any triangle, the sum of the measures of the interior angles is always 180°.

Typically, specific conjectures, once establishing their truth becomes routine, become cases from which students can see patterns that lead to generalized conjectures.

A further question to ask about generalized conjectures is ". . . always true for which set of numbers or shapes?" This question asks about the domain of the conjecture. We touched on this in Chapter 1, where we pointed out that the truth of a conjecture for a particular group of students depends on what students already know. For example, for a second grader, the conjecture

> When you multiply two numbers, the product is always greater than or equal to the two factors.

is true if the only numbers they know about are whole numbers. Once students learn about fractions, thereby expanding the set of numbers they know about, the conjecture is false.

In a previous example, a teacher had students compare the values of expressions with exponents by asking, "If we raise a fraction to some whole number power greater than one, will the result be greater or less than the fraction?" Students found that three conjectures were needed for different domains: when the fraction was greater than, equal to, or less than one.

To help students make the leap from specific to general conjectures, you can

- Have them make related specific conjectures and justify them
- Ask them to look for patterns in the specific conjectures
- Use a version of the most basic question: "What is always true, for all these conjectures you have justified?"

Pursuing Conjectures—or Not

You may wonder at times whether mathematical statements your students make really are conjectures. Mathematical statements can be used as conjectures as long as they are unjustified for some members of your class.

Conjectures that seem simple or obvious to you can be a clue as to what your students actually understand—or don't—and they can also lead to more advanced topics as well. Ms. Cooper at first considered the conjecture

> Every vertex has two numbers

to be fairly obvious. She could have interpreted this conjecture as a signal that the conventions of graphing were not obvious to all students. On further checking in with students, she may have decided to review how to graph points in a little more depth. But what if what lay behind this conjecture was not the need for remediation? Students may have wondered why only two numbers are needed to define a point. It turns out that it is only true for points in two-dimensional space. This conjecture could have led to an early discussion of three-dimensional space and how many numbers are needed to define a point there. Ms. Cooper chose not to go down this path as she was pursuing different target conjectures, as described in the section of this chapter on planning. You may, however, find that you want to pursue what at first seem like obvious conjectures but can lead to deeper understanding.

Clarifying Conjectures

Other mathematical practices or processes from state standards come into play during argumentation; in particular, clarifying conjectures is an ideal place for students to use precise language with a purpose. Ms. Cooper spent a fair amount of time clarifying conjectures with students. It's important to separate this part of conjecturing from the generation of bold conjectures.

ACCESS FOR ALL

It's important to begin by gathering all conjectures in whatever language students have access to so that participation is as broad as possible and students can learn that expressing their ideas is valued in mathematics class. Then you can engage in collectively honing the conjectures until they are mathematical statements that are both rigorous mathematically and accessible to middle schoolers. The need to communicate clearly provides motivation for understanding mathematical vocabulary and using mathematically precise language.

Visual representations can help provide access to mathematical language. For example, Ms. Cooper made a quick sketch on the board to help students use mathematical language to clarify their conjectures. Conveying the meaning of terms with displayed sketches can be an important strategy for supporting a diversity of learners, including ELLs.

Students' own questions for each other can help a great deal with clarifying. We have found that students often persist in questioning until a conjecture is well understood. Sometimes, clarifying can slip into justifying, as it did in Ms. Cooper's class. Ms. Cooper reminded students to stick with clarifying but noted that the student's contribution would be useful in the next stage, justifying.

We recommend that you do this too, particularly when you are working with beginning arguers. Keeping the parts of argumentation distinct for them provides support for learning what each part means and how to do it.

Good questions for clarifying conjectures focus on mathematical vocabulary, what is already known, and surfacing unspoken assumptions. For example,

- "Can you use more mathematically precise language for stating the pattern?"
- "What do you mean by this particular word? Is there another way to say it mathematically?"
- "What do we already know for sure?"
- "Do you think that the pattern holds for any case you could generate? If so, how can we describe that mathematically? If not, for which cases does it apply?"
- "Does everyone understand what [Student] is saying? [Student], can you say it a different way?"

Helping Students Use "If . . . Then . . ."

As they proceed through middle school, students should learn to make conjectures in "If . . . then . . ." form, although certainly not all or even most of their conjectures will take this form. "If . . . then . . ." conjectures can facilitate students making logical connections in their justifications. More about this is discussed in Chapter 6 on levels of justification. "If . . . then . . ." conjectures come in pairs: a conjecture—"If a, then b"—and its converse—"If b, then a." It's important for students to understand that these two statements are not the same, and just because one is true doesn't mean the other is true. Consider two conjectures about squares and rectangles:

If a shape is a square, then it is also a rectangle.

If a shape is a rectangle, then it is also a square.

The first is true. But the second is false because we can think of a shape that is a rectangle, satisfying the "if" part of the conjecture, but is *not* a square, so the "then" part is false. A rectangle with sides of lengths 2 units and 3 units is such a rectangle.

You can help your students start to learn to use "If . . . then . . ." in conjecturing by

- Asking, "What do we already know is true? And what is it we want to find out?"
- Asking students to formulate "If . . . then . . ." statements about nonmathematical situations, to investigate the converse and whether it is also true
- Providing "If . . . then . . ." as a sentence starter on the argumentation poster under Conjecture

ESTABLISHING NORMS

Making conjectures in mathematics will be new to many students. We suggest the following norms and warm-up games to learn the norms and help students understand what is expected of them during conjecturing.

Norms

★ **Use patterns to make statements about what will always be true.** This norm helps connect generating cases to conjecturing.

★ **Make bold conjectures about what might be true.** This norm encourages students to make the kind of conjectures that can lead to conceptual growth by pushing the envelope of students' current knowledge.

★ **Avoid judging other people's conjectures.** This norm provides space for the broadest range of conjectures.

Warm-Up Games

In addition to establishing conjecturing norms, both of the following games reinforce basic classroom norms such as mutual respect and building off each other's ideas. The game Gift Giving (e.g., Diggles, 2004) supports the conjecturing norms "avoid judging" and "make bold conjectures."

GIFT GIVING

- Players stand in pairs, facing each other.
- One player takes an imaginary gift from an imaginary closet and, through handing it over to a partner, indicates its approximate size and weight with gestures.
- The partner takes the gift, unwraps it, and says what it is.
- The first player, upon hearing what the gift is, explains why he or she got that gift for the partner.
- Players exchange roles and play again.

©iStockphoto.com/VladartDesign

In Gift Giving, students learn how to give and accept offers, which is similar to the way we expect them to make and receive conjectures. An offer is a statement or move a player makes in a game. This game relies on an improv norm that we haven't yet discussed: "Say 'yes and . . .' to offers." In other words, keep the game going instead of closing it down with "no." It won't help, in Gift Giving, for the first player to reply, "No, that wasn't what I was giving you."

> **Conjectures are like gifts, to be acknowledged and received without judgment.**

Conjectures are like gifts—they are offers that one student makes to the group. And while in the conjecturing phase, the conjecture is acknowledged and received without judgment. The analysis and discussion about its truth are left to the next stage, justification, which Ms. Thomas discusses in the vignette that follows.

The Pattern Game (e.g., Lobman & Lundquist, 2007) reinforces the norm "look for patterns" by providing an introduction to one form of pattern.

PATTERN GAME

- Players stand in a circle so that everyone can see everyone else.
- The teacher states a pattern to repeat, for example, AABB. In this example, the first and second people are A, the third and fourth are B, and the fifth starts with A again.
- Then, a player starts the pattern with a movement or a sound that others can follow, for example, saying, "Yay!" This is the first element, A, in the pattern.
- In the case of AABB, the second player will do the same thing that the first player did, repeating A ("Yay!"). Then the third player will make a new movement or sound for the group to follow, for example, waving. This is the second element, B, in the pattern. The fourth player repeats element B (waving). The fifth player starts with element A ("Yay!") again, and the pattern repeats.
- Players go around the room following the pattern until it is the teacher's turn again.
- The teacher sets a new pattern (ABC, or AAB, or ABBC), and the whole process starts again. The new pattern should start at a different place in the circle to ensure that multiple players get a chance to create, not just follow.

The Pattern Game helps students develop one way of thinking about patterns that will serve them in looking for mathematical patterns. In the game, they get concrete feedback for following a pattern. This is useful in argumentation because conjectures are often based on patterns students recognize.

VIGNETTE: NORMS FOR CONJECTURING THROUGH GIFT GIVING

Ms. Thomas had been introducing argumentation to her eighth graders, who seemed to have little experience with it. After a lesson where conjecturing stalled with only one or two conjectures made, she chose a warm-up game to begin the next class.

Ms. Thomas began class by introducing the rules for Gift Giving. She reminded students that when playing warm-up games and doing argumentation, it's important to pay careful attention, accept what others say, and build on it. She gave students 2 minutes each to give a gift to their partner and then asked a couple of students to report on what they got and why. For example,

(Continued)

(Continued)

(Continued)

Erma said she got a pencil because she never seems to have a pencil when it comes to mathematics class. Ms. Thomas then asked students to connect the game to mathematics class and a specific norm for argumentation.

Ms. Thomas: Let's see how we can connect this to the mathematics class-room. It's a fun activity, and when I first played it, I wondered, how does this connect to our mathematics classroom? Talk with your partner for 1 minute about how it is an example of talking to a respectful audience.

With this, the class connected to an important norm for any class discussion, including argumentation. Ms. Thomas went on to connect the game specifically to conjecturing.

Ms. Thomas: But this game also helps us with conjecturing in argumentation. How is that?

Ray: A conjecture is like a gift. You tell the person what you got, and then they have to, like, justify why they got it for you.

Ms. Thomas: Yes, a conjecture is like a gift. And in the game, do you reject the gift and say, "Oh, no, that's not good"? Or do you take it and respectfully receive it and work with it? That's what is going to happen as you make conjectures. You're going to think of an idea you might share with your partner or the class. Your partner or the rest of the class should try to make sense of it and respond in a way that shows they understood—and build on it. You shouldn't judge each other's conjectures but understand them and use them to help you make your own conjectures.

Notice that after students had actively engaged in the game, Ms. Thomas did quite a bit of explanation about its relationship to argumentation. You may find you can elicit this connection from students, but you should be prepared to make connections they may not see themselves.

PLANNING

In preparing for conjecturing, you'll want to consider the target conjectures to which students should get access and how to elicit them from students. Sometimes, you will provide a ready-made conjecture you want students to justify. You will also want to plan for how you and students will record conjectures in preparation for justifying them.

Target Conjectures

When planning a lesson on conjecturing, you'll need to consider both the conjectures students may make and those that you want to ensure get made—the target conjectures for the class. You can then plan for bridging between the two, in case students do not make the target conjectures spontaneously.

Ms. Cooper had planned in advance for the types of conjectures that could occur and which would be her target conjectures—the ones most directly related to the content she needed to address.

She organized the conjectures for her lesson as in Figure 3.2.

FIGURE 3.2 Classifying Conjectures

PATTERN OF COORDINATES OF VERTICES	PATTERNS SPECIFIC TO DIFFERENT TYPES OF RECTANGLES	SYMMETRY ABOUT AN AXIS
Target conjecture: The pattern of points for a rectangle parallel to the axes is (a, b), (a, d), (c, b), (c, d), where a, b, c, and d can be any numbers.	**Target conjectures:** Exactly two different numbers are needed across the coordinates when the rectangles are squares that also have a vertex of the form (a, a). Exactly three different numbers are needed across the coordinates when rectangles that are not squares have a vertex of the form (a, a). It takes exactly four different numbers to make up the coordinates of rectangles that do not have a vertex of the form (a, a).	**Target conjecture:** For rectangles that are centered around the origin, the pattern of points is (a, b), (–a, b), (a, –b), (–a, –b). **Related target conjecture:** Rectangles that are symmetric about an axis have coordinates of vertices that are opposites: Those symmetric about the y-axis have x-coordinates of opposite signs; those symmetric about the x-axis have y-coordinates of opposite signs.
Supporting conjectures: In vertices of sides parallel to the y-axis, the x-coordinates are the same. In vertices of sides parallel to the x-axis, the y-coordinates are the same.	**Supporting conjectures:** Four different numbers are not always used in the coordinates of vertices of rectangles parallel to the axes. Some rectangles have a vertex where the x-coordinate equals the y-coordinate.	

The logical connections between conjectures and the relationships to your learning goals are both important considerations as you plan for argumentation lessons. To make these considerations, you need to know what the "space" of conjectures is. This requires doing the mathematics yourself before you teach the lesson, as we suggest at the beginning of each vignette on Ms. Cooper. Once you have done the mathematics yourself and taught the lesson, you will have an expanded sense of what the conjectures are for when you teach it again.

When planning to elicit multiple conjectures, ask yourself,

- What are the basic conjectures that even struggling students can be expected to make?

- What are the target conjectures that I want to make sure everyone in the class has exposure to?
- What conjectures could connect basic and target conjectures?
- Are there conjectures that I can encourage from my more advanced students?
- What supports can I provide for my struggling students to make target conjectures?
- What are the teaching moves I can use to support my students in conjecturing?

But stay open, also, to students' conjectures that will surprise you!

Recording Conjectures (and Justifications)

Ms. Cooper recorded all conjectures on the board as students generated them in the whole-class discussion. If students are conjecturing in small groups or individually, you will want students to record their conjectures themselves. This can be done in their mathematics notebooks, or some teachers use small, blank versions of the argumentation poster on which students can record individual conjectures and, later, the justifications that go with them.

Starting With a Conjecture

Not all argumentation lessons begin with eliciting multiple conjectures. Argumentation lessons can begin with a single conjecture that the class will justify together. You can introduce conjectures that will address specific aspects of content that you know you want to cover. For example, in a lesson on geometric transformations, an eighth-grade teacher began with the conjecture

A reflection over the line $y = x$ results in the same image as a rotation of 90°.

The conjecture is based on a common misunderstanding among students. It can be used as a great opportunity to make student reasoning visible by testing it with counterexamples. As a result, students can gain insights into alternative conjectures and develop new understanding. Deliberately starting with a false conjecture is a good reasoning activity that allows students to test their ideas with examples and is a good way to develop new ideas while addressing students' misunderstandings (Campbell, Schwarz, & Windschitl, 2016).

TASKS FOR DIFFERENT GRADE LEVELS

While conjecturing follows from any of the "generating cases" tasks in the previous chapter, there are also other activities that encourage conjecturing. Some examples follow.

TASK 3.1 A TASK FOR HIGHLY SCAFFOLDED CONJECTURES, ADDRESSING THE DISTRIBUTIVE PROPERTY AND ITS ROLE IN FINDING EQUIVALENT EXPRESSIONS

_____ is an expression without parentheses that is equivalent to $2(3x + 5)$.

 Task 3.1 handout available for download at **resources.corwin.com/mathargumentation**

This task can be adapted for more advanced students by using rational numbers that are not whole numbers as the constants. There are many conjectures that students may make, depending on how well they remember or understand the distributive property from earlier grades: $6x + 5$ and $16x$ are likely incorrect possibilities for filling in the blank. For justifications, students can use visual representations or verbal expressions. We discuss the justifications in depth in Chapter 4.

TASK 3.2 FOR STUDENTS WHO ARE LEARNING TO USE SYMBOLIC
VARIABLES TO GENERALIZE AND TO FIND EQUIVALENT EXPRESSIONS,
AND ALSO TO ENCOURAGE MANY CONJECTURES

> Explore the operations on combinations of odd and even numbers. What conjectures can you make?

 Task 3.2 handout available for download at **resources.corwin.com/mathargumentation**

Students can begin this activity by defining even and odd numbers. Possible conjectures include any combination of even and odd put together with one of the four operations, some of which have already been explored in earlier chapters. Further conjectures include

The product of two odd numbers is an odd/even number.

The product of an even and an odd number is even/odd.

The sum of three odd numbers is odd/even.

A more scaffolded version of this activity can be downloaded. It asks students to first list and define even and odd numbers and then guides students through combinations of evens and odds for sums. Notice that while conjectures are scaffolded for addition, students make their own conjectures for multiplication, and the intellectual work of justifying is still left for students to do, with representations such as diagrams or algebraic symbols suggested.

ACCESS
FOR ALL

TASK 3.3 FOR STUDENTS WHO ARE LEARNING ABOUT LINEAR FUNCTIONS AND TO BUILD A CONJECTURE BASED ON A REAL-WORLD CONTEXT

Use the Desmos activity for this task: https://goo.gl/PX37U9.

What is the relationship between the unit rate of the cost per box of cereal and the steepness of the line representing the function relating the total cost to number of boxes of cereal bought?

 Task 3.3 handout available for download at **resources.corwin.com/mathargumentation**

Here are three conjectures that students might make, including a false conjecture, the last one:

The greater the unit rate, the steeper the line.

The unit rate tells you how far away from the *x*-axis the line is as you move along it.

You can't tell because the unit rate only tells you about one box of cereal, but the line is total boxes.

Looking ahead to justifying these conjectures, making a number of graphs that show the total cost in terms of number of boxes for different unit rates is helpful. Paper and pencil or digital graphing tools can be used, and we provide a Desmos activity for your use.

TASK 3.4 FOR ELICITING BOLD CONJECTURES, USING PREMADE CASES, ON THE TOPIC OF FACTORING

NUMBER	FACTORS	NUMBER OF FACTORS	PRIME FACTORIZATION
1	1	1	
2	1, 2	2	2
3	1, 3	2	3
4	1, 2, 4	3	2^2
5	1, 5	2	5
6	1, 2, 3, 6	4	$2 \cdot 3$
7	1, 7	2	7
8	1, 2, 4, 8	4	2^3
9	1, 3, 9	3	3^2
10	1, 2, 5, 10	4	$2 \cdot 5$

NUMBER	FACTORS	NUMBER OF FACTORS	PRIME FACTORIZATION
11	1, 11	2	11
12	1, 2, 3, 4, 6, 12	6	$2^2 \cdot 3$
13	1, 13	2	13
14	1, 2, 7, 14	4	$2 \cdot 7$
15	1, 3, 5, 15	4	$3 \cdot 5$
16	1, 2, 4, 8, 16	5	2^4
17	1, 17	2	17
18	1, 2, 3, 6, 9, 18	6	$2 \cdot 3^2$
19	1, 19	2	19
20	1, 2, 4, 5, 10, 20	6	$2^2 \cdot 5$

For the numbers 1–20, use the above table showing

- The factors of the number, including 1 and the number itself (for 12, they are 1, 2, 3, 4, 6, and 12)
- The number of factors there are (12 has six factors)
- The prime factorization of the number (12 = $2^2 \cdot 3$)

Task 3.4 handout available for download at **resources.corwin.com/mathargumentation**

One route students could take is conjecturing about the frequency of primes. This is a topic of current interest to mathematicians, and students can partake in it. Many other conjectures are possible. Try it! Ask yourself,

- How many factors does a prime number have?
- Which numbers have the most factors? Is there a pattern?
- What determines if a number has an odd or even number of prime factors?
- How many numbers are between every two powers of two? Do you think there is a pattern?
- What is true about the numbers that come between twin primes (which are two prime numbers that are consecutive odd numbers)? Will that always be true? The twin primes and the numbers that come between them are shaded in the chart.

The answers to these questions will be truly bold conjectures. Some may be unproven, even by professional mathematicians. Please note: The arguments for some of these conjectures are beyond what middle schoolers can be expected to do.

DIGITAL TOOLS

Online digital tools are available to help students investigate functions and equivalent expressions through graphs and symbols. Here are three sites with tools available at no cost: The National Council of Teachers of Mathematics' Illuminations website (https://illuminations.nctm.org/) provides applets, including a pan balance scale for "weighing" algebraic expressions to test their equivalence. The PhET Interactive Simulations project at the University of Colorado Boulder (https://phet.colorado.edu/) provides online simulations for mathematics and science and includes, for middle school mathematics, function graphers and motion simulations where a motion is modeled by a function. Desmos is an online graphing tool with a variety of activities available for middle school, which we have used for a couple of activities linked to in the book.

As you decide which tool to use with tasks such as those mentioned earlier, consider the role of the tool in argumentation. When investigating functions with a graphing tool, the relationship between equation and graph is demonstrated interactively: As students change the algebraic expression, they can immediately see the result of that change in the linked graph, which aids in conjecturing about the relationship. When building expressions with a virtual pan balance, students can test a conjecture about whether two expressions are equivalent, and graphs that are created as an expression is built can be used to justify why they are equivalent or not.

ACCESS FOR ALL

Each of these digital tools can be used to provide access to students for whom extensive graphing or calculation is a barrier to arguing about concepts, but they are also good for all students. Students can easily investigate a variety of expressions and functions much more quickly than if they were trying to graph with paper and pencil. Once students have used tools for a while, they may find that sketching graphs—as opposed to plotting numerous points—with paper and pencil will serve their purposes.

 WORKING TOGETHER

Here are suggestions based on this chapter for activities to do together with other teachers as you meet in grade-level groups, as a department, or as a professional learning team. They include different types of conjectures that you can use to motivate your students in learning mathematics and using precise language to communicate their ideas.

Exploration and discussion (45 min)

1. Make two to three conjectures using the activity on the topic of factoring in the Tasks section of this chapter. Use the following three questions to guide you:
   a. What are you absolutely sure of?
   b. What are you pretty sure is true?
   c. What could possibly be true? (15 min)

2. What types of conjectures will your students make? For instance, will they address a specific aspect of the content? Make conjectures that are false? (10 min)

3. Design moves to elicit false conjectures for the activity on factoring. (10 min)

4. Through conjecturing, students can gain a deeper understanding of the mathematics, learn to use precise language, and communicate their understanding to others by speaking, writing, gesturing, and using visual representations. Using the task on the topic of factoring, discuss the following questions:

   a. What is the mathematics vocabulary that can be developed through this task?

   b. Can you use more mathematically precise language for stating the conjectures you made in #1? (10 min)

Wrap-up and assignment (15 min)

5. Review the teaching moves on conjecturing in this chapter. Plan a lesson to teach before the next meeting in which students will make conjectures. Include in your plan

   • Moves that can help students build on each other's ideas

   • Target conjectures that you want to make sure are presented, including false conjectures (10 min)

6. Describe how you will teach the lesson in your class, and plan for sharing your experiences with your colleagues at the next meeting. (5 min)

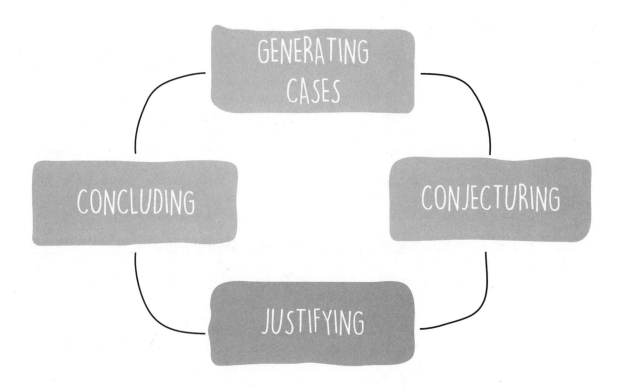

CHAPTER 4

Justifying

In this chapter, you will learn

- What justifying means in the mathematics classroom
- Through vignettes, how students can justify multiple related conjectures and critique and connect arguments
- Moves for justifying: asking why questions, prompting for justifications as a chain of reasoning, and helping students use cases
- Norms for justifying along with games to establish them
- How to plan for justifying in your lessons
- About tasks for justifying in the later middle school grades
- Through working with your colleagues, to develop new moves for helping students in developing their mathematical authority

WHAT DOES IT MEAN TO JUSTIFY?

When students justify a conjecture, they are presenting the reasons why it is true or false. A justification can be thought of as a logically connected chain of statements, beginning with what students know to be true and ending with a decision on the truth or falsity of a conjecture—even though justifying, as it unfolds in real time, may not follow that chain in order from beginning to end. A good justification convinces others; it goes beyond a personal exploration of an idea.

This chapter addresses two aspects of justification: eliciting justifications and helping students connect and critique justifications. We address each through its own vignette and discuss the set of teaching moves that supports it.

Justification engages students' prior knowledge and also helps students build new concepts. In the two vignettes in this chapter, students build on their knowledge of number and geometry from elementary school to develop new understandings of variables and coordinate geometry.

> My challenge is to constantly remind students about what's important with mathematical thinking versus procedural thinking—and having the students communicate their ideas to other people, using the mathematics.
> —Eighth-grade teacher

VIGNETTE: JUSTIFYING MULTIPLE CONJECTURES

In the following vignette, Ms. Cooper and her students are engaged in the justification task as set out in the Rectangle Coordinates activity.

> Justify one conjecture based on what you know about coordinate geometry. Remember, your conjecture may turn out to be true or false.

Before you read the vignette for this chapter, choose two of the conjectures students made in the vignette for Chapter 3 and justify them using what you know about geometry, both with and without coordinates. Think about how the two conjectures are related by their justifications.

In this lesson, Ms. Cooper helped her students justify some of the conjectures they had made the day before in a whole-class discussion. She was prepared to help students represent their patterns using symbolic variables, an important step in early middle school algebra, as reflected in her target conjecture.

> Ms. Cooper began the day's lesson by reminding students of a few conjectures they had previously created that she wanted the class to justify.
>
> Ms. Cooper: We have a lot of good conjectures—so many that we can't discuss them all today. Here are three that I'd like us to try to justify today—and remember, that means finding out why they are true or false.

She displayed three conjectures on the board alongside a diagram she had created of four rectangles in the coordinate plane.

1. For the vertices of vertical sides of the rectangles, there were two coordinates the same.

2. For the horizontal sides, there are two *x*-coordinates that are the same.

3. When you write down the coordinates of the vertices of any rectangle (parallel to the axes), you will use four different numbers that are in different orders for the different coordinate pairs.

She had deliberately included one false conjecture, Number 2.

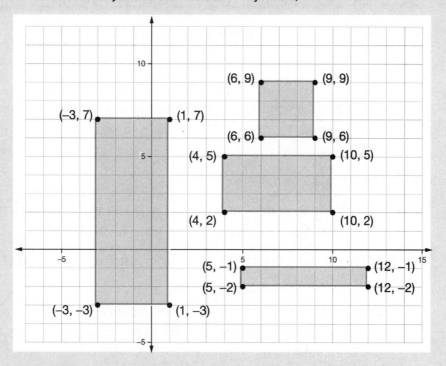

She wrote the first conjecture on the argumentation poster.

Conjecture   [. . . Always . . .]

For the vertices of vertical sides of the rectangles, there were two coordinates the same.

Ms. Cooper: Let's start with the first conjecture. Does anyone have an idea why that is true, no matter what vertical side we pick, or why it's false?

Ying: Look at that rectangle in the middle. On the vertical side, the numbers 4 match up.

(Continued)

Ms. Cooper: Where in the coordinates do you see the number 4?

Ying: It's the *x*-coordinate for both corners.

Ms. Cooper revoiced the argument that Ying made to make it more precise.

Ms. Cooper: So to say it a little more precisely, in the rectangle, the *x*-coordinates of the vertices on one of the vertical sides are both 4. Is that what you mean?

Ying: Yes, I think so.

Ms. Cooper: OK, are we done? Have we justified the conjecture?

Bailey (pointing to all four rectangles): Yes, we see it right there.

Camden: The conjecture is supposed to work for *all* vertical sides. Maybe there could be one where it doesn't work.

Ms. Cooper: That's a good insight, Camden, that we need a justification that works for *any* vertical side of a rectangle. Who thinks they know?

Several students raised their hands.

Dale: Well, no matter what vertical side you pick, in any of the rectangles, the *x*-coordinate is going to be the same.

Ms. Cooper: Yes, that's our conjecture. But why is it that the *x*-coordinates have to be the same?

Emma: It doesn't have to be just those rectangles. It works for any of them. Because they will always be a certain amount away from the *y*-axis.

Ms. Cooper wrote on the poster, replacing "a certain amount" with the words "the same distance":

The vertices of a vertical side of the rectangle are always the same distance from the *y*-axis.

Ms. Cooper, realizing Emma's part of the justification did not yet address the *x*-coordinates, pressed further:

Ms. Cooper: How does that relate to the *x*-coordinates?

She called on Lavar, whose hand was not raised but whose group had made this conjecture the day before.

Lavar: The *x*-coordinates tell you how far you are from the *y*-axis. That means for vertical sides, the *x*-coordinates all the way up and down are the same, not just the vertices.

Ms. Cooper asked if everyone understood what Lavar was saying. Monika had a question:

Monika: I just don't see why the *x*-coordinates are always the same.

Ms. Cooper: Lavar, can you explain your thinking in more detail to Monika and the class?

Lavar: Look, on the *y*-axis, all the *x*-coordinates are 0. That makes sense, right? So now take a vertical side. If you go over from the *y*-axis to the line, you always go over the same amount. Because both lines are vertical. That means they are parallel—always the same distance apart is how I learned it.

Ms. Cooper asked Monika and the class if they were satisfied that the conjecture was true, and all thumbs went up. She noted also that Lavar had brought in the term *parallel* to make his explanation.

Getting thumbs up from around the room, she recorded what Lavar had said.

Ms. Cooper: Are we done? Are you convinced?

All thumbs went up again.

The completed poster looked like this:

MATHEMATICAL ARGUMENTATION

Conclusion   [. . . Always . . .]

> For the vertices of vertical sides of the rectangles, there were two coordinates the same.

Justification   [How do we know?]

> The vertices of a vertical side of the rectangle are always the same distance from the *y*-axis.
> This is because the vertical side is parallel to the *y*-axis.
> The *x*-coordinates tell you the distance from the *y*-axis.
> Because of that, the *x*-coordinates are the same for all the points on the side.
> So the *x*-coordinates of the vertices on a vertical side have to be the same.

Conclusion   [True or False]

> True

(Continued)

(*Continued*)

Ms. Cooper: Let's summarize. We wanted to show that the conjecture is true. We knew that the sides are vertical. How did that lead us to know the *x*-coordinates of the vertices on that side would be the same?

Mimi: Well, we said that vertical sides are always the same distance from the *y*-axis. And it's the *x*-coordinate of the points that tells us the distance of the points from the *y*-axis. That means the *x*-coordinates of vertices on a vertical side are always equal.

Monika: And Lavar said the vertical lines are parallel. That's what convinced me.

Ms. Cooper: We said this justification convinced us all, so we can fill in the conclusion.

She wrote on the poster:

Conclusion: True

Ms. Cooper: Notice this justification started with what we already know, and built from there, to establish whether the conjecture is true or false. That's what a good justification for a conjecture that is true should do—you should be convinced every step of the way. Every step followed from the one before it. Now let's take the second conjecture.

Most students agreed with Conjecture 2, but a couple thought that it was false.

Rowan (pointing to a horizontal side of one of the rectangles): It can't be true, because it says the same thing as for vertical sides, but it's about horizontal sides. Look at that horizontal side; the *x*-coordinates are not the same, but the *y*-coordinates are.

Rowan had established the conjecture as false by giving one counterexample. When students agreed on that, Ms. Cooper asked if they had a new conjecture. They reformulated the conjecture as

Conjecture 2B: For all the points on a horizontal side, the *y*-coordinates are the same.

She called on some students who had struggled with argumentation, knowing they could follow the same structure as for the first justification and make a valid justification themselves. Using the same ideas about distance from the axis, these students successfully supported the conjecture.

Ms. Cooper: OK, now that we have those two pieces in place, let's look at our last conjecture.

She wrote it on an argumentation poster:

Conjecture   [. . . Always . . .]

> When you write down the coordinates of the vertices of any rectangle (parallel to the axes), you will use the four different numbers that are in different orders for the different coordinate pairs.

One student had an insight right away.

Sachi: They are in different orders but they follow a pattern.

Ms. Cooper: Let's see if we can use what we have already established as true, and see what that pattern is.

Preparing to help them use variables, on a grid with no axes, she drew a rectangle and a set of parentheses at every vertex, ready for the students to fill in, without allowing them to simply read off particular numbers. She reminded students that they had just established that the x-coordinates of the vertices on the vertical side are the same and asked how they could represent that with variables. A student said that they could name them both with the same letter, and Ms. Cooper labeled the two coordinates a.

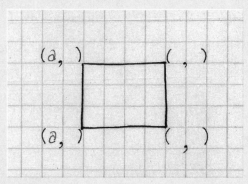

Students went on to label b, c, and d. They agreed that these labels followed from the two statements they had already justified.

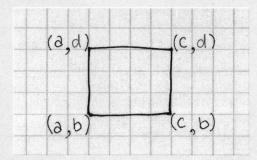

(Continued)

(*Continued*)

She went back to the conjecture.

> Ms. Cooper: So what does all this tell us about this conjecture? We really clarified it some more, and justified it at the same time.

> Aliyah: There is a pattern and it goes (*a, b*), (*c, b*), (*a, d*), (*c, d*) because of what we said before.

> Ms. Cooper: How do you know this pattern works for all rectangles like these?

> Malcolm: The *a*s have to be the same because they are on a vertical side. Same for the *c*s.

> Nadia: It's the same for the *b*s and *d*s, but horizontal.

> Lavar: You could make that rectangle big or small or move it around; *a, b, c,* and *d* are still there.

Ms. Cooper has been filling in the argumentation poster as they went along:

MATHEMATICAL ARGUMENTATION

Conjecture   **[. . . Always . . .]**

> For any rectangle with sides parallel to the axes, the vertices follow the pattern (a, b), (c, b), (a, d), (c, d).

Justification   **[How do we know?]**

> The vertical sides have the same *x*-coordinates, so that makes the *a*s and *c*s the same.
> The horizontal sides have the same *y*-coordinates, so that makes the *b*s and *d*s the same.

Conclusion   **[True or False]**

> True

She asked if the class was convinced that the pattern held for all rectangles, and they were.

> Ms. Cooper: OK, great. We used different variables to represent different numbers. But let's think for a minute. Is it possible for *a* and *c* to be equal? What would happen if they were the same number?

Aliyah:	The rectangle would squish together and wouldn't be a rectangle at all.
Ms. Cooper:	And we could make the same argument for *b* and *d*. What about *a* and *b*? Could they be equal?
Sachi:	They are in one of the rectangles we started with.
Ms. Cooper:	Right, so when we use different variables, we are saying they don't have to be the same—but they could be. Here is where algebra really helped us think about all rectangles. That's what coordinate geometry is all about—algebra helps us think about geometry and geometry helps us think about algebra. How else did algebra help us—in making our arguments go beyond the cases we had made?
Sachi:	It wasn't just about the rectangles on the display anymore; it was about all the rectangles that could be made.

In this lesson, students had solidified their knowledge of properties of line segments that are parallel to the axes and had extended their use of numeric representations to algebraic representations. Ms. Cooper was then prepared, in a subsequent lesson, to have students address the standard about how to find the length of line segments parallel to the axes.

TEACHING MOVES FOR ELICITING JUSTIFICATIONS

Ms. Cooper used a number of teaching moves that you can use yourself when you help students justify their conjectures. You'll need moves for broadening access to and explaining justifying, different ways to ask "why," and moves that follow up on the "why" questions.

Broadening Access Through Justifying

Ms. Cooper broadened access to mathematical authority with several techniques specific to justifying. She used students' own conjectures as the starting places for justification, even when her target conjecture was a bit beyond those conjectures. She called on a student whose hand was not raised, whom she suspected had an additional contribution to the argument, based on his prior work. Additionally, she continually focused students on their own mathematical authority by frequently asking if everyone agreed and encouraging respectful disagreement.

ACCESS FOR ALL

Explaining What Justifying Means

The process of helping students understand what justification is and what constitutes a good justification is ongoing. It's important to comment on qualities of good justifications as they arise in the classroom. Not only will students' justifications improve over the school year, but you also begin to set the stage for students' engagement in more formal proofs in high school.

Ms. Cooper let students know what a good justification does by highlighting qualities of a justification that students had just made. She said that their "justification started with what we already know and built from there." Sometimes, she summarized their argumentation like that, but other times, she elicited from students the qualities of a convincing or less-than-convincing justification, for example, that a finite set of cases is not enough to establish that a generalized conjecture is true.

A Basic Supply of "Why" Questions

Using good questioning techniques takes practice. As we explained in Chapter 1, if argumentation is relatively new for you, we recommend that you start by simply asking "why" during a lesson that you always teach. Ms. Cooper asked two "why" questions in a row to help students build a justification based on basics that everyone was sure of. First, she asked, "But why is it that the x-coordinates have to be the same?" And then, when students hadn't quite made the connection to x-coordinates, she probed their developing argument with "Why are the points on the side all a certain distance away from the y-axis? How does that relate to the x-coordinates?"

In this vignette, Ms. Cooper used some variations on "why" as well. You can extend your "why" questions into other formats to elicit further discussion and justification from students.

How Do We Know It Is True?

As discussed in Chapter 1, this is a central question to use often. Its purpose is to focus students on the idea of truth as opposed to a solution being right or wrong. It asks students to both draw on their own reasoning and convince others. It can be invoked whenever a statement is made that hasn't as yet been established as true. Eventually, you can expect students to incorporate the question into their own mathematical thinking, when they say things like "I think . . . because . . . " without being prompted.

What Makes You Think So?

This extension of a "why" question is good for eliciting tentative justifications from students who are just beginning to justify or who are moving into uncharted mathematical territory. Ms. Cooper's question for the same purpose was "Does anyone have an idea . . . ?" She sensed that students weren't quite sure where to start, because it was early in the year, and she was still establishing norms about taking risks in mathematics class. She had also seen that some students were weak in their facility with coordinate geometry when they were generating cases, and so she thought that they might not be very confident in making arguments using it. So she asked for students' initial ideas to get the argument started.

Why Does It Make Sense That . . . ?

Students who are just beginning to make arguments often simply repeat the conjecture, as Dale did in the vignette, or provide a procedure instead of a generalized reason. Asking "Why does it make sense that . . . ?" pushes students past simply demonstrating steps for how to arrive at an answer and focuses on reasoning (Cioe, King, Ostien, Pansa, & Staples, 2015). For example, instead of simply asking why, asking "Why does it make sense that the pattern goes up by 4 each time? Can you show that in the diagram?" will help students focus on reasoning about relationships (p. 487). You can read the Cioe et al. article for further ideas about how to support student justification, particularly in algebra. Additionally, we have more to say about students' levels of justification and how to help them advance in Chapter 6.

How Will You Convince Us That This Is True?

This question explicitly requests that the speaker provide a convincing justification for the classroom audience. It comes from *Introduction to Reasoning and Proof: Grades 6–8* (Thompson & Schultz-Ferrell, 2008), a good resource for learning more about argumentation in the middle grades.

What Comes After "Why"

Sometimes, students need a little help in answering a "why" question. Ms. Cooper used a couple of different types of probing questions:

- Eliciting the properties of a mathematical object that are useful to the justification at hand. For example, Ms. Cooper asked, "How does thinking about the x-coordinates help us?"
- Asking for relevant special cases. For example, Ms. Cooper elicited squares as special cases. Special cases can be used as counterexamples to help prove that a conjecture is false, or as a set of cases supporting another conjecture.
- Asking students to respond to each other's questions. As with Lavar in the vignette, answering another student's question brought a new concept into play.

Building a Common Chain of Reasoning

Justifications can be thought of as a chain of reasoning, starting with what is known or assumed, and leading to a conclusion about the truth of the conjecture. You can help students build that chain together, as opposed to a set of separate arguments from various students. Ways to do so include

- Introducing the hand signal of putting fist over fist to make a building motion to indicate that students want to continue an argument that another student has started
- Introducing the norm of careful listening, as discussed in Chapter 1 and reinforced by many of the warm-up activities we recommend
- Using the prompt "Because of that?" to help create a chain prompt a new student's contribution to it

Be prepared, though—a collaboratively created justification may not follow a logical chain of reasoning as it unfolds; in that case, there is time to summarize it logically in the concluding phase, discussed in Chapter 7.

Helping Students With Precision

Most state standards emphasize attending to precision in trying to communicate mathematically. As students build justifications out of mathematical statements, you will find natural opportunities to help students use mathematical language and symbols precisely. Precise language is both mathematically correct and meaningful to its user and his or her audience.

Revoicing is an important way to help students move toward precision in language. In revoicing, you restate what students have said but refine it mathematically. Ms. Cooper revoiced "the same amount apart" as "same distance," an important concept in developing understanding of signed numbers on a number line. The context of the justification helped students comprehend the more precise statement.

Precision also comes into play in definitions. For example, students often talk about rectangles as having "two long sides and two short sides," based on common rectangles they have seen. By that definition, squares are not rectangles. But most mathematical definitions of rectangle include all four-sided figures with four right angles. These definitions include squares as rectangles.

In summary, you can help students understand and use precision by

- Allowing students to use their own language for mathematical ideas, but connecting that to more accepted mathematical language through revoicing.
- Encouraging students to use clear definitions and properties that follow from them in their justifications.

Supporting the Use of Cases in Justifying

The use of cases is a theme that runs throughout this book. Starting in Chapter 1, counterexamples were introduced. "Generating cases" is its own phase in argumentation, the topic of Chapter 2. Patterns across cases suggest conjectures, as discussed in Chapter 3. In Chapter 6, on levels of justification, we address the ways in which students use cases as their sophistication in argumentation grows. In this chapter, we discuss some basic distinctions in how cases are used in justifying.

As discussed in Chapter 1, counterexamples are single cases that disprove a conjecture. Building on what happened in the vignette, Ms. Cooper's students went on to consider the conjecture that four different numbers are always needed to write the coordinates of the vertices of a rectangle. One student offered the counterexample of a rectangle with one vertex where the x-coordinate and y-coordinate were equal and therefore used only three numbers, showing the conjecture to be false. Even the revised conjecture that just three or four numbers are needed turned out to be false, since a square with the points (1, 1) and (4, 4) requires only two numbers across all the coordinates of its vertices.

Case-based arguments of why a conjecture is *true* are a different matter. A finite number of cases is not sufficient to establish that a generalized conjecture is true. However, cases—even one case—can suggest a generalized justification that works for all cases, even an infinite number. Ms. Cooper's students started justifying by offering one rectangle as an example for which the conjecture

For the vertical sides of the rectangles, there were two coordinates the same

was true. She asked her students whether this was an adequate argument by asking, "Are we done?" One student cited the four rectangles on the public display as further justification. Another student objected, and Ms. Cooper pointed out that they wanted an argument that worked for *any* rectangle—reinforcing the generalized language in the conjecture. This began their generalized justification.

You can prompt students to use cases in justifying by asking,

- "Can you find a counterexample—one example that shows us that the conjecture is false?"
- "Can we use these cases or examples to think about what happens in every case?"
- "Can we represent all of these [numeric] cases with a variable?"

VIGNETTE: CRITIQUING AND CONNECTING ARGUMENTS

There's more to say about justifying in terms of supporting students arguing together and, in particular, critiquing the reasoning of others. Let's look at a vignette from Mr. Flores's class, where students were justifying a conjecture based on a task about the distributive property from Chapter 3:

$6x + 10$ is an expression without parentheses that is equivalent to $2(3x + 5)$.

Before you read the vignette, try justifying this conjecture for yourself, without simply citing the distributive property but by explaining why it works. In mathematics, a property is a characteristic that something has because of its inherent structure. So it is with the distributive property: It follows from the relationship between multiplication and addition—that multiplication can be expressed as repeated addition.

In the following vignette, Mr. Flores guided students through two justifications of why $6x + 10$ is equivalent to $2(3x + 5)$ and then asked students to connect those two justifications.

> After students worked for a few minutes in small groups, Mr. Flores called the class together and asked for one group to share its justification.
>
> Mariana: I think it is $6x + 10$ because that's how we learned to do it last year.
>
> Nash: But that's not a justification. It doesn't tell you why.
>
> Mr. Flores: So how might you show why $2(3x + 5)$ is equivalent to $6x + 10$? Work in your groups for a few minutes. Try a couple of examples. Then try at least one of these: a justification in words, a justification using a diagram, or a justification using symbols.
>
> After about 10 minutes, Mr. Flores called on a group of students who had made a diagram to come to the front of the class and show what they had done.

(Continued)

(Continued)

Rayhan: The parentheses tell us that this means 2 times $3x + 5$. We know that multiplication can be shown on a rectangle, so we drew this:

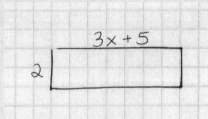

Rayhan: Then we figured out what is really in the rectangle. Each one of the 3 xs gets multiplied by 2, so that is $3x + 3x$. And the 5 gets multiplied by 2, so that is 10.

Sara: So add it up, that is $6x + 10$.

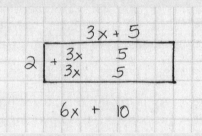

Mr. Flores invited critiques of the argument.

Mr. Flores: Does anyone disagree with their justification?

Nash: But why isn't it $6x + 5$? That's how you would multiply.

Priya: Well, just try that—say x is 4. It doesn't work.

Nash: I don't get it.

Priya: The first one, if you make x equal to 4, it means $2(3 \cdot 4 + 5)$. Do what is in the parentheses first. For example, 12 plus 5 is 17. And 17 times 2 is 34. Does everyone agree?

All thumbs went up.

Priya:	Now take $6x + 5$. If you substitute in 4, you get $24 + 5$, which is 29. They aren't equal.
Mr. Flores:	And because of that?
Nash:	The sentence says they are equal no matter what x is. But when x is 4, it doesn't work.
Omar:	I have a different question. Their diagram makes it look like x is smaller than 5. What if x is larger than 5? Will it still work?
Sara:	I could make the xs longer. It would work the same way because everything still has to be doubled.

Mr. Flores asked if anyone else had questions for Rayhan and Sara. When no one did, he said that their argument based on a diagram was convincing. Mr. Flores then called on a group that had done its justification using symbols.

Mr. Flores:	Bea and Cesar, how did you justify the conjecture?
Bea:	You don't really need to draw it out. We said that parentheses mean multiplication. But multiplying by two is just adding a thing twice. So 2 times the amount $3x$ plus 5 means $3x$ plus 5 plus $3x$ plus 5. If you combine like terms, then it's $6x$ plus 10. So we got the same.

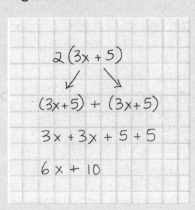

Nash:	What's x?
Bea:	It doesn't matter what x is; the mathematics still works.
Mr. Flores:	Is there anyone else who has questions about this argument? No? Then this is another good justification of our conjecture.

Mr. Flores then turned to connecting the two justifications.

Mr. Flores:	Now, can you explain how the two justifications show the same thing, even though one is a diagram and one is in symbols?

(Continued)

(Continued)

Cesar:	They both use doubling the amount in the parentheses.
Mr. Flores:	That's important to remember: The symbols inside the parentheses represent an amount—an amount that changes as *x* changes—for example, *x* can be greater or less than 10, as we saw. How was doubling shown in each justification?
Fabia:	Once it was drawing the same picture twice, and in the other, it was adding to itself.
Mr. Flores:	Does it matter that we used the number 2? These arguments all use doubling, which is what you do when you multiply by 2.
Eli:	I don't think so. Because if it's 3, that's tripling; if it's 4, that's quadrupling, and so on. It's just multiplication.
Mr. Flores:	Eli is saying that we could replace 2 in the expression with any number. Does anyone disagree?

After giving some time for students to respond, he concluded,

Mr. Flores:	No, so really, we have established something more general than just the original conjecture. Now, use this idea to write two other equations that show how the distributive property works with variables.

TEACHING MOVES FOR CRITIQUING AND CONNECTING ARGUMENTS

The vignette illustrates some ways to help students critique the reasoning of others, and we discuss them. In addition, we discuss ways to make connections between arguments.

ACCESS FOR ALL

Broadening Access Through Critiquing and Connecting

Although his techniques differed from Ms. Cooper's, Mr. Flores also broadened access to mathematical authority. His frequent move, asking if anyone disagreed, clearly encouraged respectful disagreement and for students to voice their insights. He let another student address a misapplication of the distributive property through a counterexample. Additionally, he treated students' pictorial and symbolic arguments equally, helping students see how they complemented each other.

Supporting Students in Critiquing Arguments

When students critique the reasoning of others, they respond directly to a justification that other students have made. For example, two students asked questions following Rayhan and Sara's presentation of their argument using diagrams. One of them, Nash, challenged the justification by offering her own conjecture. Rayhan and Sara explained why her conjecture was false with a counterexample. The other student, Omar, questioned the generality of Rayhan and Sara's argument, and they explained why their visual justification would work for values of x less than or greater than 15. Both of these methods—offering a counter-conjecture and questioning whether a conjecture is true in all cases—are ways in which students may challenge each other's arguments. Mr. Flores encouraged these sorts of critiques by asking if anyone disagreed with the argument, rather than asking about agreement. This gave students a sense of permission to voice their critiques. In Mr. Flores's class, in contrast to Ms. Cooper's, the classroom community's decision that a conjecture was true or false was indicated by lack of disagreement.

You can help students critique arguments by

- "Opening the floor" for disagreement
- Using sentence starters such as "I agree/disagree with . . . because . . . " to help beginning arguers structure their critiques
- Probing: "Which part of her argument do you agree with? Disagree with?"

Connecting Justifications

Mr. Flores asked students to connect the two representations from the two justifications his students had presented. The concept of doubling, a very accessible version of multiplication, arose in both justifications and allowed students to see the parallels between the symbolic representations (multiply by 2 or add the amount to itself) and visual representations (draw the bars twice). In making these connections, students had the opportunity to deepen their understanding of the distributive property and how it relies on the relationship between addition and multiplication.

Finally, Mr. Flores prompted the opportunity to connect both justifications to a more generalized version of the distributive property by asking if there was anything special about doubling as multiplication.

You can help students connect arguments by

- Explicitly asking students to compare two arguments—for example,
  - "How do these two justifications show the same thing?"
  - "How does the visual representation in this justification help us understand the numbers/algebraic symbols in this other justification?"
- Probing: "How does this part of your justification relate to the one you just saw/heard?"
- Asking for a more generalized conjecture based on two or more justifications of a specific conjecture. (We distinguished these two types of conjectures in Chapter 3.)

ESTABLISHING NORMS

Norms for justifying help students understand some fundamental principles about the nature of justification and about creating justifications together. Warm-up games help you establish these norms.

Norms

- ★ **Look for reasons why a conjecture is true or false.** This very basic norm establishes the need for justification in the first place.
- ★ **Build off of other people's ideas.** Students may start out justifying by giving reasons that are disconnected from each other; one student presents his or her reasons why a conjecture is true, and then another student presents a completely different idea. While there will always be room for new ideas, to really establish argumentation as a social practice, as a common search for the truth, students need to learn to connect their contribution to prior contributions.
- ★ **Try to convince others of your ideas, but keep in mind that you could be wrong—and that's OK.** Argumentation is about being convincing to others, but it's not about being personally right or wrong. Just as conjectures that turn out to be false lead to learning, so do justifications that aren't completely valid. Students can think of ideas for how to correct a justification or get new insights from seeing that it doesn't support the conjecture and why.

A Warm-Up Game

A useful metaphor for mathematical argumentation is storytelling. Similar to stories, mathematical arguments have a flow that starts with a conjecture, builds during the justification, and ends in a conclusion. The warm-up game Story Spine (Adams, 2007) helps build the metaphor and helps students see how the argumentation model is similar to a story. The spine is a nine-line form that gives a framework for telling stories in a group. This warm-up game makes the process of building off each other's ideas very explicit and concrete. Students must listen to each other and understand the previous students' sentences in order to make a sensible story.

STORY SPINE

Write on chart paper so that the whole class can see the following sentence starters:

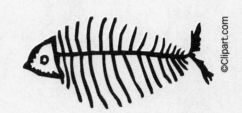

Once upon a time . . .
Every day . . .
But one day . . .

Because of that . . .
Because of that . . .
Because of that . . .
Until finally . . .
Ever since that day . . .
The moral of the story is . . .

- Players stand in a circle. One player begins by reading the first sentence starter and finishing it as a single, complete sentence. (It should not be a run-on sentence.)
- The next player then reads the next sentence starter and finishes that sentence in order to start making a story.
- After eight people, the story has been told and the ninth person adds a moral to the story that sums up the main theme.
- The class should play two or three times, continuing around the circle until all students have had a chance to make a sentence.

Variation: The first time, we recommend playing this with the whole class, even if not all students get to participate. You can discuss the norms about listening to each other and speaking loudly enough for others to hear.

The sentence starters provide a structure that lays out important ideas that build a narrative. This structure enables players to make many stories. After playing the game, you can help students see how Story Spine is to stories as our argumentation model is to arguments. For example, "Once upon a time . . ." is similar to the conjecture that opens an argument. The three sentence starters "Because of that . . ." can be used to build a mathematical justification. "Until finally . . ." is similar to the conclusion. And, "The moral of the story . . ." is similar to what you do as a teacher to explain what has been learned through the argument.

After playing the game, you can help students see that using an argumentation framework such as conjecture–justify–conclude, like the Story Spine, helps them build an argument that will make sense. You can remind students of the purpose of each statement in the story, for example, the logical connections between statements connected by "because of that . . . "

PLANNING

In preparing for justifying, you'll want to consider in advance possible student conjectures and organize those conjectures, considering logical connections and their relationship to your learning goals for students.

Thinking Through Possible Justifications

It's essential to work out in advance at least one justification for each of the conjectures you plan to have students address. It can be challenging to follow students' justifications as they emerge in classroom discourse. You increase your chances of making sense of what students are saying if you have at least one justification in mind; even if it is different from the ones students offer, it may lend insight into what they are saying.

Organizing Conjectures for Justification

It often happens, as it did for Ms. Cooper, that generating cases and conjecturing occur in one lesson, and then justifying is the focus of the next lesson, all based on the same activity. This provides you the opportunity to organize students' conjectures to serve your instructional purposes during the lesson on justifying. You will want to consider your chosen target conjectures, which are closely aligned with your learning goals, as described in Chapter 3. Also consider the conjectures students may first make that lead to your target conjectures. Then, there are different ways to organize students' conjectures for justifying. For example, you may want to organize the conjectures logically so that each conjecture, once justified, can be used to support the next conjecture. Or, you may want to organize conjectures by complexity or sophistication of the mathematics needed to justify them. Sometimes, those two methods result in the same ordering.

To teach the justification lesson based on Rectangle Coordinates, Ms. Cooper built on what she did in the prior lesson on conjecturing to organize a chain of conjectures. These conjectures included both the conjectures students had made the day before and her target conjecture. She saw that if students worked out justifications for two of the more basic conjectures about vertical and horizontal sides, the third, encapsulating both conjectures, would follow more easily. Figure 4.1 shows a diagram of this in her planning book.

FIGURE 4.1 Teacher Writes Out and Organizes Students' Conjectures

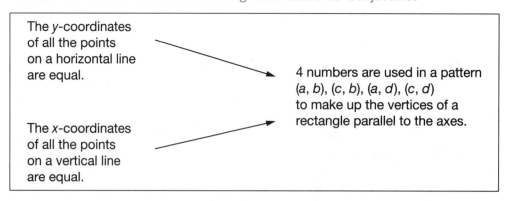

With this order, she would address her instructional goal for students to apply and extend their use of numeric representations to developing algebraic representations to represent the overall pattern in the coordinates of rectangles with sides parallel to the axes.

In summary, when planning for justification, ask yourself:

- What are the learning goals for the lessons? How do the conjectures students have already made help students achieve those goals?
- What are justifications for the conjectures we will address?
- How should I order the conjectures for justification so that the justifications build on each other?
- Which teaching moves will I need to support students in justifying?

TASKS FOR UPPER GRADE LEVELS

Because justifying builds on conjecturing, which builds on generating cases, any of the tasks in Chapters 2 and 3 can be used for justifying. We reexamine some of those here and provide example justifications.

The vignettes in this chapter have focused on earlier middle grades standards. Here are some tasks aligned with later-year standards:

FOR STUDENTS WHO ARE LEARNING ABOUT PROPORTIONAL RELATIONSHIPS AND TO BUILD A CONJECTURE BASED ON A REAL-WORLD CONTEXT

> What is the relationship between the unit rate of the cost per box of cereal and the steepness of the line representing the function relating the total cost to number of boxes of cereal bought?

 Note: This task is the same as Task 3.3. Task 3.3 handout available for download at **resources.corwin.com/mathargumentation**

Use the Desmos activity for this task: https://goo.gl/PX37U9.

This task was introduced in Chapter 3. One important conjecture students can make that prepares them for slope is

> The greater the unit rate, the steeper the line.

Here, cases for justification are useful. With graph paper and pencil, students can draw graphs for different unit costs, or you can have them use the Desmos activity we have prepared. A good justification would be

> The unit cost tells you how much you will pay for one box of cereal. The unit rate is represented by a point $(1, t)$ on the line graph. We know it is a line because "per" defines a proportional relationship. The greater t is, the higher the line is "aimed," so the line will be steeper.

This pre-slope language builds on an intuitive notion of what steepness is and prepares students to consider the change in x (represented by 1) and the change in y (represented by t) separately, which establishes the idea of slope.

TASK 4.1 FOR STUDENTS WHO HAVE LEARNED ABOUT THE PYTHAGOREAN THEOREM AND ARE READY TO APPLY IT

> What is the length of the line segment with endpoints (0, 3) and (4, 0)? How can you find the length of any line segment, if you know the coordinates of its endpoints? Justify your method.

 Task 4.1 handout available for download at **resources.corwin.com/mathargumentation**

Use the GeoGebra activities for this task:

Part 1: https://goo.gl/yPYe7Q

The conjecture will be some version of what can be stated most generally as,

The distance between (a, b) and (c, d) is the square root of $[(a - c)^2 + (b - d)^2]$.

The Pythagorean theorem figures prominently in the justification of this conjecture. To help students see the connection, you can ask them to find the length of the line segment with endpoints $(0, 3)$ and $(4, 0)$. The Pythagorean triple may be fairly obvious for the triangle formed when the endpoints of the segments lie on axes. They can then generalize from this insight.

FOR STUDENTS WHO ARE STUDYING TRANSFORMATIONAL GEOMETRY

Plot the vertices of a triangle. Reflect it over the x-axis. What patterns do you see in the coordinates of the vertices of the triangle and its image?

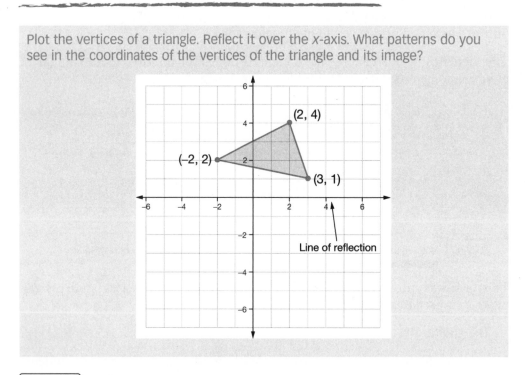

online resources — Note: This task is the same as Task 2.4. Task 2.4 handout available for download at **resources.corwin.com/mathargumentation**

This task was introduced in Chapter 2 for generating cases. Students can generate cases using vertices of triangles. Then students' conjecture would be

The image of (a, b) reflected over the x-axis is $(a, -b)$.

The justification would depend on how reflections have been defined. Here is one definition of reflection that is appropriate for middle schoolers: "A reflection is a transformation of the plane that flips each point of the plane across a line to a point that is the same distance away from the line." Then a good justification is the following:

We know from the definition that the reflection takes a point to another point the same distance from the x-axis. The x-axis is where $y = 0$. We know that points that are opposites are the same distance from 0. So the y-coordinate of the image must be the opposite of the y-coordinate in the original point. Writing that algebraically, if the original point is (a, b), then the image is $(a, -b)$.

DIGITAL TOOLS

In addition to the GeoGebra files provided for this chapter, there are other free websites that have interactive activities for helping middle school students learn transformational geometry. One of these, the National Library of Virtual Manipulatives (a well-established National Science Foundation–funded project, http://nlvm.usu.edu./en/nav/vlibrary.html), has apps for many mathematics topics and grade levels. It has three apps on transformational geometry: one each on translation, reflection, and rotation. These apps are fairly open-ended tools in which students can make a shape and play with its image under one transformation.

WORKING TOGETHER

Here are activities you can do with other teachers in grade-level meetings, in department meetings, or as a professional learning team, to explore tasks, moves, and planning for justification.

Exploration and discussion (50 min)

1. Review the tasks in this chapter. In grade-level groups, make a conjecture based on one of the tasks and write a justification appropriate to your grade level. (20 min)

2. Outline a lesson for justification based on the task. Be sure to include possible student conjectures for the lesson. (15 min)

3. Review the teaching moves in the chapter. Come up with your own moves—questions or prompts—that you can use in your lesson for the following purposes:

   - To get students started in justifying a conjecture, including one that may be false
   - To help students in developing their mathematical authority
   - To support students in using mathematical precision in communicating their ideas
   - To move students beyond justifying with a single example or case
   - To guide students in comparing justifications and critiquing the reasoning of others (15 min)

Wrap-up and assignment (10 min)

4. Select a lesson to teach before the next meeting that includes justifying based on one of the tasks in this chapter. After you teach the lesson, write out a few notes that you can use at the next meeting to help you reflect with your colleagues: What moves did you use? For what purposes? How did your next steps (homework or exit tasks, for example) follow from the justifications students made?

NOTES

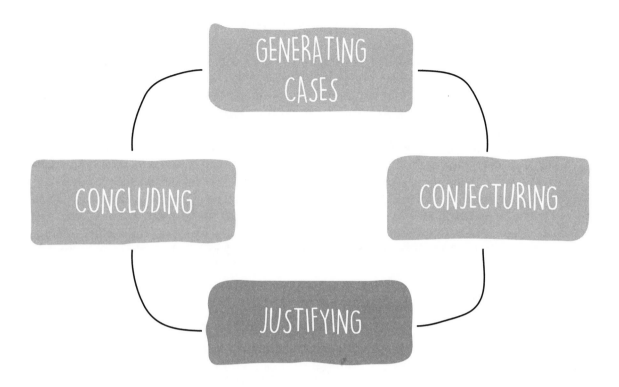

CHAPTER 5

Representations in Justifications

In this chapter, you will learn

- What representations are and how they enhance justifying
- Through vignettes, how gestures, diagrams, other visual representations, and dynamic software support students, including those receiving special education services and ELLs, in conjecturing and justifying
- Moves for justifying that help students make sense of symbolic representations and bridge to more formal mathematics
- How dynamic digital tools support students' learning
- Norms for helping students pay attention to representations such as gesture
- How to plan for students to create visual justifications, including specific tools you will need to provide
- Tasks for justifying that encourage students to use representations
- In meetings with your colleagues, the affordances of using particular representations for different groups of students

WHAT ARE REPRESENTATIONS?

As mathematics teachers, we are used to thinking about "the big three" representations—graphs, tables, and equations—and these are certainly important in justifications. Other representations for argumentation include sketched diagrams, dynamic digital representations, verbal representations, and even students' gestures. This expanded set of representations opens up a variety of means of engaging with and expressing mathematics content. By allowing students to use these representations in argumentation, you can gain a better understanding of all students' thinking—thinking that may otherwise be difficult to access. For example, students can use diagrams to express the same algebraic ideas that symbols can represent, as we saw in Mr. Flores's vignette in Chapter 4. In the broadest sense, representations can be used to "portray, clarify, or extend a mathematical idea" (NCTM, 2000, p. 68). We provide more detail on how that can happen in this chapter.

ACCESS FOR ALL

You will find that compelling representations can come from many different students. Dynamic digital representations and, in contrast, the use of students' gestures in the air are particularly important for ELLs or for students whose disabilities make it difficult for them to use language. The two vignettes that follow provide examples of such students contributing to argumentation through the use of such representations, though their use is certainly not limited to these populations.

VIGNETTE: VISUAL REPRESENTATIONS FOSTER PARTICIPATION

Ms. García had been using dynamic geometry software for her argumentation lessons with a class of seventh-grade students, including some who were receiving special education services. Her learning goals for students were to classify triangles based on their angles and sides and to make arguments about the relationships between angles in a triangle.

Here is the handout for the activity she was doing with students. Try writing and justifying statements about the types of the triangles yourself, with the help of the GeoGebra activity, before you read the vignette.

1. Based on your exploration, for each pair in the table, check off whether it is a shape that is *possible* or *impossible*.

	Scalene	**Equilateral**	**Isosceles**
Acute	☐ *Possible* ☐ *Impossible*	☐ *Possible* ☐ *Impossible*	☐ *Possible* ☐ *Impossible*
Right	☐ *Possible* ☐ *Impossible*	☐ *Possible* ☐ *Impossible*	☐ *Possible* ☐ *Impossible*
Obtuse	☐ *Possible* ☐ *Impossible*	☐ *Possible* ☐ *Impossible*	☐ *Possible* ☐ *Impossible*

2. In the table above, write sentences that describe the relationships between each of the triangle types. The words *some*, *all*, or *none* can help us describe relationships clearly. Use the following sentence structures if you need help.

    It is impossible to make _____ from _____.

    Some _____ are also _____.

    All _____ are _____.

3. Because we have not yet justified these statements, they are all still conjectures at this point. Choose one that you would like to justify.

4. Justify your conjecture using what you know about the definitions and properties of different types of triangles.

Use the GeoGebra activity for this task: https://goo.gl/xaFDyx. Created with GeoGebra (www.geogebra.org).

 Task 5.1 handout available for download at **resources.corwin.com/mathargumentation**

Ms. García had her students generate conjectures in pairs using a premade triangle on which they could show the measure of any side or angle. They could also move the vertices around to form different kinds of triangles. When her students were done conjecturing, she brought the class together to justify one of their conjectures. She wrote the conjecture on the argumentation poster:

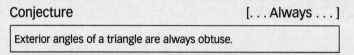

Conjecture   [. . . Always . . .]

Exterior angles of a triangle are always obtuse.

Then, she shared on the whole-class display a triangle with two exterior angles marked at vertex C.

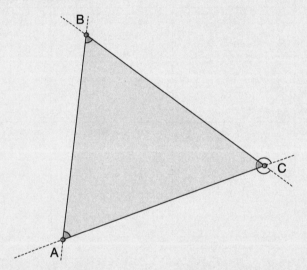

Ms. García wanted to confirm that all the students understood the conjecture, so she asked what they thought *obtuse* meant and how they could check it. Students used a piece of paper as proxy for a 90° angle, to visually check whether the exterior angle at C was obtuse. Then, Ms. García returned to justifying the conjecture. She knew her students could easily come up with a counterexample by moving the vertices of the triangle.

Ms. García: Kendra, do you want to come up and see if you can manipulate the angle so that it's not obtuse anymore? Grab vertex C and see if you can manipulate vertex C so that the exterior angle isn't obtuse anymore.

By asking if students *could* come up with a counterexample, she left open the question of whether the conjecture was true or false. Kendra tried to make a counterexample by moving vertex C slowly to the left.

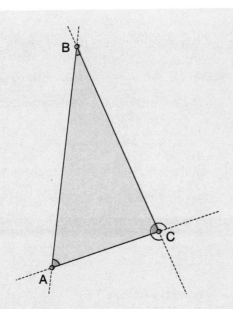

Ms. García: Still looks obtuse to me.

Kendra's inability to come up with a counterexample didn't prove the conjecture was true. So Ms. García further pursued a justification.

Ms. García: Can anybody else manipulate it?

Jesse: I'll do it.

Jesse, a student who rarely participated or spoke in class, came to the front of the class and quickly dragged vertex A to the right of vertex C. This made the interior angle at C an obtuse angle and the exterior angle at C an acute angle. Jesse raised his hands in a winning gesture to convey his confidence in having demonstrated that the conjecture was false.

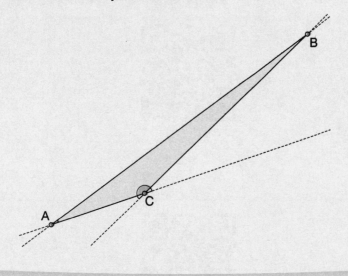

(Continued)

(Continued)

Ms. García then filled out the argumentation poster, adding to the justification:

[Conclusion]   [True or False]

> We were able to make a triangle with an exterior angle that
> was less than 90°. So, the conjecture is false.

She explained that one counterexample is sufficient evidence to disprove this conjecture.

VIGNETTE: GESTURES ENABLE A UNIQUE CONTRIBUTION

Here's the activity for the following vignette. Try it out for yourself to get a sense of what the conjectures in the vignette refer to.

TASK 5.2 SQUASHING AND STRETCHING A PARALLELOGRAM: WHAT HAPPENS TO THE AREA?

Use the two GeoGebra files to explore what happens to parallelograms when the side lengths are or are not fixed:

Part 1: https://goo.gl/Wkh6Bw

Part 2: https://goo.gl/iEx8uU

Inspired by a conversation with students in the previous class, Mr. Lima posed the following argument to his bilingual students, including ELLs (Lara-Meloy & Barros, 2000):

> The area of a rectangle and the area of a parallelogram are the same because if you collapse a rectangle into a parallelogram, the area doesn't change.

Mr. Lima presented a student argument that was not yet mathematically precise so that students would have the opportunity to provide refinements to it. After a few minutes in which students worked in pairs, Mr. Lima asked for volunteers, and several students offered competing justifications.

Mr. Lima:	Let's start with you two, Fatu and Lena.
Fatu:	The area will be the same.
Mr. Lima:	Okay. How do you know?
Fatu:	Because we drew it and the formulas for area are the same.
Mr. Lima:	Why don't you bring your notebook to the document camera and you can walk us through your logic?
Lena:	See, the base and the height are the same, so the area is the same.

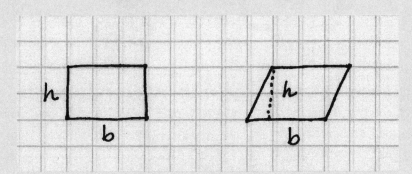

Tiago jumped up:	But this is not what the argument said. It says what will happen as the rectangle flattens.
Fatu:	Yeah. You can see that the parallelogram is flatter. You can make it really long, and it will still be the same area.

(*Continued*)

(*Continued*)

She drew another longish parallelogram with the same height.

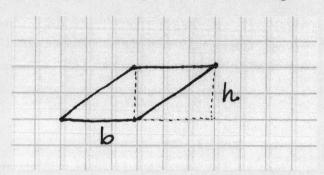

Fatu: You see the triangle here (pointing to the triangle formed at the left side of the parallelogram), that's the same one as here (pointing to the right of the parallelogram). So, the area stays the same.

Tiago: But if a rectangle flattens . . . it's like a box. If you go all the way, it will have no area.

At this point, Tiago extended his arms and bent them at the elbow, simulating the sides of a rectangle. He then simultaneously pivoted both arms at the elbow until his forearms were making a line and the right hand was touching his left elbow.

With this gesture, Tiago expressed that the sides of the parallelogram were fixed and that flattening a rectangle was more like flattening a box, changing the angle. At this point, Mr. Lima stepped in to help students develop language to express Tiago's idea.

Mr. Lima: Do folks understand what Tiago is trying to say?

Some students said yes, but others said no.

Mr. Lima: I want you to take a minute to talk to your partner to discuss what Tiago just said. See if you can make a diagram to help us understand what he said.

After a few minutes, Mr. Lima asked if anyone wanted to explain what Tiago was gesturing in words.

Anika:	We agree with Tiago, and we disagree with Fatu and Lena. We made this drawing before, when we were discussing.
Mr. Lima:	Remember to explain what Tiago said using his arms.
Anika:	I was saying that it's the same thing as what we said. If you squoosh a rectangle, the height is getting less, until there's no more height.

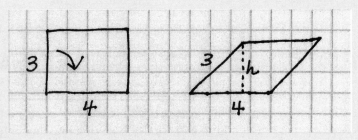

Mr. Lima:	How does this affect the area?
Anika:	If there's no more height, then there's no more area.
Mr. Lima:	Fatu, Lena, what do you think? Are you convinced?
Lena:	I think we had a drawing that had the same height in the rectangle and the parallelogram. But Tiago is saying that the sides are the same, not the height. If you keep going, and you squoosh it all the way, the height gets smaller, and you end up with the top line of the parallelogram on top of the bottom line, so there is no area.

As this vignette shows, gesture and diagrams can both play an important role in ELLs' participation in argumentation. Tiago's gesture had been critical in helping students understand Tiago's new conjecture they eventually articulated:

> As one pair of interior angles of a parallelogram gets smaller, the area of the parallelogram becomes less.

The gesture had also been part of a justification for it. Mr. Lima had students articulate that gesture and justification in words.

TEACHING MOVES

Helping students use representations in their justifications requires a set of teaching moves including both eliciting and providing representations and the tools to make them.

Show How You Know

"Show how you know" is really another form of the "why" questions presented in Chapter 4. It invites students to use various representations to demonstrate their reasoning. It focuses them on what they know for themselves as well as on their classmates as an audience for their representations. You can deploy this question when students seem to have run out of words, or when they are having difficulty with symbolic representations.

Providing Appropriate Tools

Follow-up moves to "show how you know" include giving students appropriate tools for making representations. For example, students can use 3D manipulatives such as linking cubes or models of geometric solids as the basis for justifications about number, algebra, and geometry. Sometimes, a simple sketch with paper and pencil is sufficient to visually represent important ideas. For example, to represent an essential property of proportional relationships, student can use a pair of more-or-less perpendicular axes with a sketch of a line through the origin. On the other hand, a carefully plotted graph on graph paper or with a digital tool can reveal the relationship between rows in a ratio table and help students justify ways to make new entries in the table.

Introducing a representation can provide access to a new concept or a new level of generalization. For example, Ms. Cooper, in Chapter 4, provided a rectangle with no coordinates labeled and no axes displayed as a way to prompt a generalized, algebraic conjecture. Students more comfortable with algebra may develop this generalized symbolic representation on their own. You can also encourage students to independently choose and develop their own representations. For example, students may use a table from which to reason while others may prefer insights they can get from a graph.

Using Visual Representations to Bridge to the Formal

Visual representations can be a critical part of a justification when the formal mathematics is a bit beyond students' reach. For example, in studying transformations, middle schoolers may work on justifying the conjecture that reflection over the line $y = x$ "switches" the order of the coordinates. A formal justification in high school requires substitution of y for x and x for y in the algebraic representation of any point (x, y) and its image, as well as the use of theorems (previously justified conjectures) about the congruence of triangles. But middle schoolers could graph the line $y = x$ and justify the conjecture by folding the graph paper on the line and demonstrating how any point and its image are related as they match up on the folded paper.

At the other end of the spectrum, visual representations can bridge between elementary school– and middle school–level reasoning. You may, for example, want to keep a set of linking cubes—a common elementary school manipulative—in your classroom. Students working on the conjecture

   The sum of two consecutive numbers is odd

could lay side by side two stacks of cubes representing 4 and 5 and see that the stacks form pairs with one left over. This can lead to the insight that they can keep adding pairs of cubes to the stacks to represent any two consecutive numbers. The visual of the pairs helps students make sense of the symbolic representation of the sum of two consecutive integers, $x + (x + 1)$, and how it simplifies to $2x + 1$.

Including Students' Gestures

ACCESS FOR ALL

Gestures—nonverbal body movements that coincide with speech—support students who are learning new concepts to communicate their thinking and mathematical ideas. We all make gestures as we talk, and it's important to pay attention to how they contribute to mathematical justifications. For bilinguals or ELLs, gestures are an important way to show what they mean, even when they don't have the right words in English or mathematical language. Gestures can provide students with a way to emphasize an idea or to provide more precision to a seemingly more informal statement.

In mathematics, gestures are especially useful, as they are an additional visual representation, a means to embody abstract mathematical concepts (Moschkovich, 2002). For example, Tiago's gesture enabled him to contribute something important for the whole class, revealing for all students how to visualize what collapsing could mean, geometrically, and how to reason about the related areas.

You can be purposeful about the gestures you make yourself: "Be obvious," as we say in the norms section of this chapter. The game in that section is particularly important for highlighting gesture.

Summary of Moves

In summary, moves to use with representations include

- Providing a specific tool that will move an argument in a productive direction—for example, a blank table handout to save precious class time and help students see relationships numerically
- Prompting for new representations—for example, "Could you try graphing the rows in the table as points? What shape will they make?"
- Encouraging students to manipulate representations in dynamic tools using premade files or web pages
- Asking students to generalize based on a visual representation—for example, "How does this picture apply to any odd number?"
- Asking students to watch one another as they speak, paying attention to how students use gestures to show their thinking

USING DYNAMIC DIGITAL TOOLS

Dynamic digital tools are such an important category of representations with special properties that we devote an entire section to two frequently used types.

Dynamic Geometry Tools

Jesse is an example of a student who receives special education services who was able to use dynamic representations to aid in his participation in class. But many of Ms. García's students benefited from the use of dynamic geometry files in which triangles were premade for them to manipulate and measure. Those who lacked fine motor dexterity did not have to tediously draw the triangles needed for the activity. With one move, Jesse was able to show his idea for a counterexample for the conjecture in question. His demonstration was so vivid that he didn't need to use words to explain it. While it may not always be easy to gather enough working computers to use in your classroom, it can be worth it for reaching a broad range of students.

The power in this kind of software comes from students being able to directly manipulate geometric objects and test out an infinite number of cases where some features of the objects stay the same during manipulation. Students can observe patterns that are mathematically meaningful, leading to geometric conjectures. They can look for evidence to see if the patterns are the same regardless of the changes resulting from dragging, which then provides insights for justifications that can be formulated in the language of geometry. They might come across a counterexample during their exploration that challenges their initial conjecture. Geometric objects in these environments have to be constructed in ways that will enable patterns to be revealed—it's often best to make them in advance for middle schoolers. Ms. García, for example, in the vignette on visual representations, used three intersecting lines to form a triangle. The vertices were then "hot spots" that students could drag while keeping the triangle intact. Students could then measure the angles on which they wanted to focus. Fortunately, you don't have to be an expert in dynamic geometry file creation in order to use either tool; both GeoGebra and The Geometer's Sketchpad® have active user-community websites where other teachers and curriculum developers post files that you can use. Additionally, we have provided a GeoGebra activity for the activity in this vignette. A good resource for research on dynamic geometry is a book by Sinclair, Pimm, and Skelin (2012). In addition to this research, books such as *Rethinking Proof* by Michael De Villiers (1999) are useful sources of tasks for argumentation in geometry.

Dynamic Function Tools

Another kind of digital tool provides dynamically linked representations of functions. An example is MathWorlds software, with a simulation of motion (distance vs. time) that is dynamically linked to a graph, a table, and an equation. This software is freely available through the University of Massachusetts's Kaput Center, along with curricula. Similar commercial products are available. A common starting place with the software is having students directly manipulate two line graphs representing the change in position of two objects over time and then having them predict which motion will be faster. Students quickly catch on to the relationship between the steepness of a graph and its speed, which is quantified in the table and equation (which can also be manipulated). The more formal definition of slope directly follows from quantifying steepness. In the case of unit pricing, the graph represents total cost versus number of items purchased, and the unit rate is the slope of the graph. There is a well-established body of research showing the advantage of such software for conceptual learning, including articles by Roschelle et al. (2010).

ESTABLISHING NORMS

In this section, we provide norms for using representations in justifications and a warm-up game that will help you establish those norms with your students.

Norms

To encourage students to consider different representations whenever they are justifying, you can establish two norms and call upon them often:

- ★ **Show it a different way. Make a drawing, table, or graph.** Simply telling students to make a new representation may not be sufficient for them to know what to do. This norm includes specific ideas about what it means to represent a problem or justification in a different way.
- ★ **Be obvious.** This norm encourages students to try to make their contributions clear to the rest of the class, calling on whatever representations will help make their mathematical point, including gesture.

A Warm-Up Game

In the game Magic Clay (e.g., Warren, 2008), students create an object through gesture and motion. Gestures are an alternative representation—because students don't use words in order to explain or convey the meaning of the object. And, similar to Off the Shelf in Chapter 2, this game also reinforces the generative nature of argumentation and collaboration. The more obvious students are, the clearer it is to their audience what object they are envisioning.

MAGIC CLAY

©iStockphoto.com/tttuna

This clay is all pretend!

Everyone stands in a circle and is able to see everyone else.

- The first player grabs a piece of imaginary clay from a pile in the center of the group and molds it into something (for example, an umbrella, a kitty cat, a ring, or a basketball). The teacher can do this first and be very obvious so that the group can see what he or she is doing and recognize the thing he or she is making.

(Continued)

(*Continued*)

- Players should be able to guess what the clay has become. If players cannot guess, this is an opportunity to state that one of the norms of this game is to be very obvious.
- When the first player is done, she or he passes the clay to another player, who must make something new out of it. The class guesses again.
- Keep passing the clay until most have taken a turn.

Hint: Players can grab more clay from the pile to make a larger piece or take chunks away from the clay they receive to make it a smaller piece.

Variation: For large classes, arrange smaller groups of players.

Be sure to discuss with students how gestures allowed students to "see" an object that the player was intending to make and that this was done without words, graphs, or other forms of expression. Highlight that they should look at each other while making mathematical arguments so they can see how gestures contribute. Remind them to be obvious when making arguments—to take on the responsibility to be explicit in conveying their ideas as clearly as possible and with any tools necessary: words, gestures, and diagrams.

PLANNING

In preparing for helping students use representations in justifying, you'll want to consider the role of the representations, whether to suggest representations or let students choose their own, and the materials needed.

Considering How Representations Support Justifications

You can consider in advance how different representations may come into play in justifications. Sometimes, these will be the "big three," but be sure to consider how alternative representations may be used.

Ask yourself: Do you know that certain students are inclined to use one representation or another? For example, Mr. Flores, in Chapter 4, knew that some students were comfortable with area models to represent multiplication, based on prior work they had done. He felt confident that at least one of those students would use that visual representation, but he was prepared to suggest it to a small group if needed. He wanted to use it in the whole-class discussion as a way to ground students' more symbolic work.

Also ask yourself: How does a certain representation contribute to a justification? For example, a teacher of eighth graders, Ms. Khan, was preparing to help students justify the conjecture that all proportional relationships are linear, but not all linear relationships are proportional. She thought through two arguments based on different representations. In terms of equations, all proportional relationships between two quantities x and y can be represented by $y = kx$ for some number k. And all linear relationships can be represented by $y = kx + c$ for some numbers k and c. So any equation $y = kx$ can be rewritten as $y = kx + 0$, which is a linear equation where $c = 0$. But any time that c is not equal to 0, then that linear equation is not proportional. In terms of line graphs, proportional relationships can be defined as those represented by lines that go through the origin, which is a subset of all lines, and any line, whether it goes through the origin or not, represents a linear relationship. So, depending on how students initially define proportional or linear, based on cases of each that they examine, the argument about their relationship is different. Ms. Khan knew that students might mix representations in their justifications, and sorting them out in advance would aid her in helping students keep them straight.

Materials for Representations

While you may want to sometimes provide a specific representation for a specific task, it's often a good idea to allow students to choose their own representations. Students should have access to materials such as graph paper, rulers, protractors, compasses, templates for tables, colored pencils and markers, 3D manipulatives, and poster paper. If at all possible, provide access to dynamic digital tools—which could be through a few handheld devices, tablets, or laptops that groups can share. Think in advance about what kinds of materials support the most powerful representations for the tasks your students will work on.

You will want to pay special attention to planning for how students will share representations with the whole class. Sharing representations is as important as making representations, because justifications are intended to convince others. We have seen the power of inviting a student to the whiteboard in establishing shared mathematical authority. Consider also handheld whiteboards that students can use to show their sketches or examples to each other. With posters, students get to organize their casual sketches and notes for public display. A document camera, in contrast to a poster, allows students to directly share their small-group work with the class and can save time. A computer- or tablet-based display enables students to show dynamic representations in motion.

In summary, the kind of display you choose is based on what is available to you and for what purpose it will be used. It's worth it to push for access to all of the following:

- Individual whiteboards or a document camera where students display quickly produced representations including sketched diagrams
- A projector or interactive whiteboard where students can share digital representations and manipulations
- Posters that enable students to refine their representations before presenting them—you can use our argumentation posters, available from links in Chapters 1 and 2, or have students create their own on large paper

online resources | Argumentation posters available for download at **resources.corwin.com/mathargumentation**

Graph paper is not just for graphing. It can help students organize tables, perform operations on multidigit numbers and make diagrams with elements aligned. We recommend that students use graph paper notebooks for all their mathematics work.

TASKS FOR DIFFERENT GRADE LEVELS

TASK 5.3 FOR STUDENTS LEARNING ABOUT DIVISION OF FRACTIONS

A shortcut way to describe how to divide fractions is "invert and multiply." What could this mean? Why does it work?

online resources → Task 5.3 handout available for download at **resources.corwin.com/mathargumentation**

Students may work on a justification based on a particular division of fractions problem. You can encourage them to choose easy fractions such as $\frac{1}{2}$ as a divisor, to make it easy to see what is going on. You may want to help them with representations. You can provide a verbal representation such as "I have $2\frac{1}{2}$ yards of ribbon. Each fancy bow I want to make takes $\frac{1}{2}$ of a yard of ribbon. How many fancy bows can I make?" You can then ask them to make their own word problems that help us understand how division of fractions should work. Figure 5.1 shows a visual representation of division of two fractions.

FIGURE 5.1 Division of Fractions

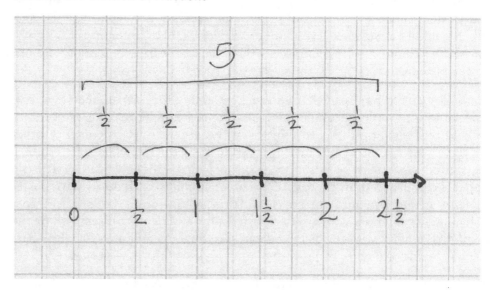

For a video of a teacher working with students making arguments about fraction division, see *Connecting Mathematical Ideas* (Boaler & Humphreys, 2005), where students show why "1 divided by $\frac{2}{3}$ is $1\frac{1}{2}$."

TASK 5.4 FOR STUDENTS LEARNING ABOUT PROPORTIONAL RELATIONSHIPS

Proportional relationships can be represented by equations of the form $y = kx$, where k can be replaced by any number. What are the properties of the table of a proportional relationship? Why? What are the properties of the graph of a proportional relationship? Why?

online resources → Task 5.4 handout available for download at **resources.corwin.com/mathargumentation**

Use the Desmos activity for this task: https://goo.gl/afyy4V.

Encourage students to make graphs and tables that are based on equations of the form $y = kx$. Their tries could look like this:

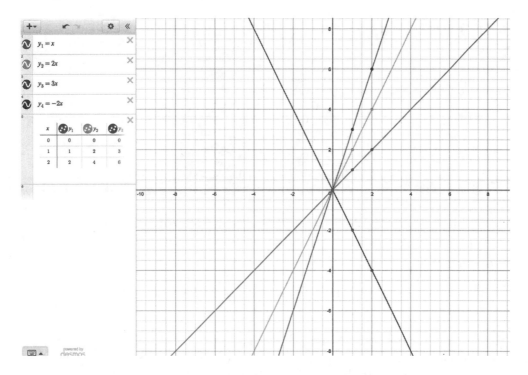

Here is a justification for how tables and graphs show proportional relationships:

I know the equation means that every time you increase x by 1, y increases by some number like 3. This means that the table for a proportional relationship will "jump" by 3 in the y column when the numbers in the x column are 1, 2, 3, and so on. The graph of a proportional relationship is always a straight line because the points are linked by the same size stair steps if you go up by 1s. The graph of any proportional relationship goes through the origin, because when x is 0, y has to be 0 too, since any number multiplied by 0 equals 0.

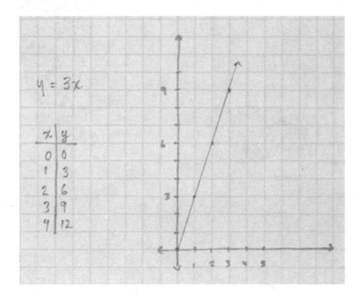

Use the GeoGebra activity for this task: https://goo.gl/yZQtH7. Created with GeoGebra (www.geogebra.org).

What is the relationship between the volumes of a cone and a cylinder with the same radius and the same height? How do you know?

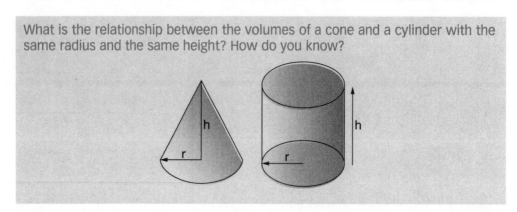

online resources Task 5.5 handout available for download at **resources.corwin.com/mathargumentation**

A conjecture is:

> By looking at the shapes, I estimate that three of the cones can fit into the cylinder.

The formula for the volume of a cylinder is the product of the area of the circular base and the height. The formula for the volume of a cone is one third of the volume of a cylinder that has the same base area and height. In symbols,

$$\text{The volume of a cylinder: } \pi \bullet r^2 \bullet h$$

$$\text{The volume of a cone: } \tfrac{1}{3}\pi \bullet r^2 \bullet h$$

To help students justify their conjecture about this relationship, be sure to provide 3D models for their use, preferably hollow ones with an open end. Encourage them to design a test to see if their conjecture was right. At the middle school

FIGURE 5.2 Empirically Justifying the Relationship Between Volumes

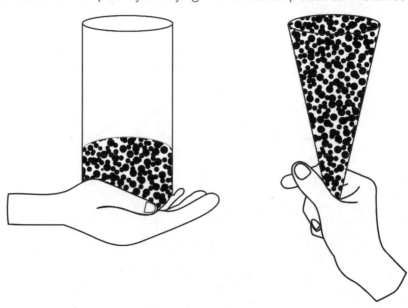

level, the argument will be empirical—it will be based on what they can observe from such a demonstration (as in Figure 5.2). This test can serve to underpin a deductive proof when they study calculus.

In this meeting with your professional learning community, grade-level team, or department, you will examine through vignettes how different students use visual representations in the classroom. You also get to come up with ideas for teaching moves that you can use to support students in using visual representations to justify their conjectures and communicate their understanding to others.

Exploration and discussion (40 min)

1. Consider the affordances of certain representations for different groups of students. Think about how the representations in the vignettes could be used beyond the groups discussed in the vignettes. Share your ideas with a partner. (20 min)

2. Working with your team, quickly review the teaching moves for encouraging students to use representations in justifying. Then make a list of your own moves—questions and prompts—and the tools you will use in your own classroom for the following purposes:

   • Help students focus both on what they know for themselves and on their classmates as an audience. Be sure to include the tools you will use.

   • Provide a bridge to more formal mathematical expressions or concepts. Include the tools you will use. (20 min)

Wrap-up and assignment (20 min)

3. Outline a lesson that you will teach during the next week, including

   • A task that lends itself to using visual representations for justifying a conjecture

   • Moves that you will use for supporting the students in using visual representations

   • How you will encourage students to create and use their own representations

   • Opportunities you will provide for students to share their representations with the whole class

At the next meeting, report on your experiences teaching this lesson.

NOTES

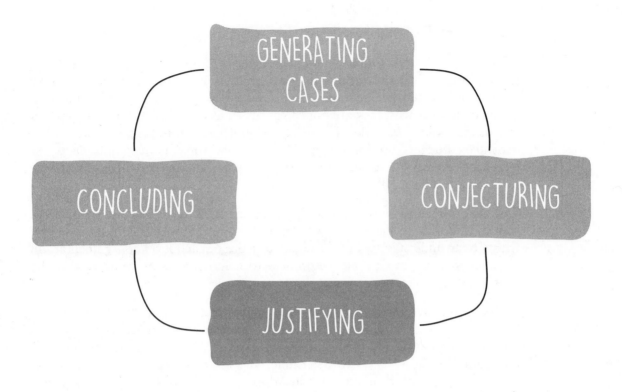

CHAPTER 6

Levels of Justification

In this chapter, you will learn

- Four levels for characterizing students' justifications
- Moves for helping students progress through levels of justification
- Through working with your colleagues to review students' work, how to assess levels of justification

FOUR LEVELS OF JUSTIFICATION

Students' justifications come in various levels of sophistication. In this chapter, we describe categorizations that will guide you in helping students make more sophisticated justifications over time. These aren't really developmental levels; in other words, they are not categorizations of *students*, but categories of the *justifications* they make. The level of sophistication of the justification will depend on students' understandings of a particular topic. For example, a student might make a Level 1 justification about geometry, but make a Level 2 justification about the parity of numbers.

We have drawn on a body of foundational research (Harel & Sowder, 1998, 2007; Knuth, Choppin, & Bieda, 2009; Lannin, Ellis, & Elliott, 2011) to provide you with a clear understanding of the characteristics of each level of argumentation. We also provide practical advice on how to help students move to making the next level of justification.

In line with the research, we categorize justification into four levels (see Figure 6.1). At *Level 0*, students don't do much more than repeat the conjecture or cite an authority ("My teacher told me . . ."). At *Level 1*, students will rely on a handful of cases to determine whether a conjecture is true or false. Students can get stuck making justifications at this level without the help of a skilled teacher, but research has shown that even elementary students are capable of understanding the limitation of using just a few cases. Your support is key in challenging students to move to making *Level 2* justifications in which they build on cases to create partially generalized arguments. *Level 3* justifications are fully generalized and sound, drawing on definitions and previously established mathematical truths.

FIGURE 6.1 The Four Levels of Justification

Level 0   No justification
Repeating the conjecture or citing authorities such as the teacher

Level 1   Cases only
Providing one or more cases only as a justification, even for a general conjecture

Level 2   Partial generalization
Explaining cases in a way that supports generalization

Level 3   Fully generalized and logically sound
Using logic to carefully build arguments statement by statement

LEVEL 0: NO JUSTIFICATION

There are three common ways in which students fail to justify at all, especially when you first start asking why. It's not that they are *not* answering the question posed; it's just that the answer isn't a *mathematical* justification.

Repeating the Conjecture or Providing a New Conjecture

In the following student work samples, students are asked to justify why they agree or disagree with the statement that

> The sum of any two consecutive numbers is always odd.

Figure 6.2 is a typical response when students think that the statement is so obvious that there is no need for justification, or when they don't know what "justify" means. In Figure 6.3, the student offers an alternative or opposing conjecture without justifying that the original one is false. This kind of counterconjecture (as opposed to a counterexample) is not an argument for why the conjecture is false but instead is another statement that remains to be justified.

FIGURE 6.2 Restating the Conjecture

Question 1

Sonia says, "The sum of any two consecutive numbers is always odd."

Consecutive numbers are whole numbers that are next to each other in order.

Do you agree with Sonia?

You just have to add the numbers & they end up odd.

FIGURE 6.3 Making a Counterconjecture

Question 5

Darius says, "If I multiply any two consecutive numbers I will get an odd number."

Remember, consecutive numbers are whole numbers that come one right after the other.

Do you agree with Darius?

No, because the key word is multiply, and if you multiply a consecutive number you'll get an even number.

Relying on External Authority

Your students may have years of training in relying on the teacher to judge correctness when it comes to mathematics. Or, "because it says so in the book" may seem good enough to them. While you, as the teacher, are the final arbiter of mathematical truth in the classroom, and the mathematics textbook is an important reference, neither of these citations constitutes a mathematical justification. The responses in Figure 6.4 are typical responses for relying on external authority when students are asked to justify.

FIGURE 6.4 Relying on External Authority

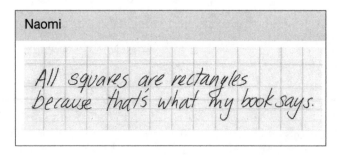

Naomi

All squares are rectangles because that's what my book says.

Malik

Faster speeds are always represented by steeper slopes. I know this because my teacher says so.

Using Visual Evidence Alone

Students may use "because they look like it" as an argument based on visual representations instead of thinking about the mathematical properties of the objects the figures represent. In Figure 6.5, Kingston says that two triangles are similar simply because they look alike in a diagram, without analyzing properties of the triangles that the diagram reveals (such as side lengths). Zwena draws on the definition of parallel, but the argument is based only on the part of the lines that show in the picture instead of the geometric objects, which are two lines that *do* intersect at some point not on the page.

FIGURE 6.5 Justifying Using Only Visual Evidence

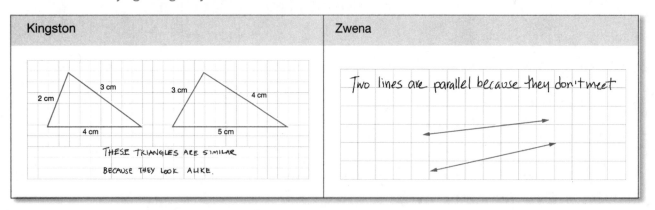

Kingston

2 cm   3 cm   3 cm   4 cm

4 cm   5 cm

THESE TRIANGLES ARE SIMILAR BECAUSE THEY LOOK ALIKE.

Zwena

Two lines are parallel because they don't meet

LEVEL 1: CASE-BASED JUSTIFICATIONS

Level 1 justifications depend on only a few cases. The cases may confirm the conjecture, but the argument ends with the cases—no generalizations, explanatory language, or algebraic symbols accompany it. Also, students working at this level may not understand the role of a single counterexample in establishing that a conjecture is false. Level 1 justifications are quite common for elementary and middle grade students, and even for high school and college students who haven't been taught explicitly about argumentation in earlier grades. Yet, students of all ages can be supported in going beyond Level 1, as we outline in the teaching moves section that follows. There are several varieties of Level 1 justifications you'll want to notice.

Believing That More Cases Makes for a Better Argument

When you ask students to go beyond arguing based on a couple of cases, they may simply produce more cases. In nonmathematical arguments, it's often true that more evidence is better. In mathematics, however, more evidence does not equal the better argument. The following dialogue illustrates a Level 1 justification for a conjecture that

The sum of the measures of the interior angles for any triangle is 180°.

Abby:	I checked two triangles with angles of 30°, 60°, 90° and 45°, 45°, 90°. Both add up to 180°, so I think the conjecture is true.
Roneesha:	I looked at another triangle with different angles from yours, of 20°, 20°, 140°. They also add up to 180°, so the conjecture must be true.

Students may have more conviction that a conjecture must be true based on more cases. But having more cases does not guarantee that a conjecture is *always* true, because it is not possible to examine an infinite number of cases. Students making Level 1 justifications may also believe that they need more than one counterexample to show that a conjecture is false, in order to be extra convincing.

Randomly Choosing Cases

Students' choices of cases at this level may not help them generalize. The cases may appear random. Or students may inadvertently choose only cases from which patterns are difficult or impossible to see.

Figure 6.6 illustrates random choices two students made to show that the sum of any two consecutive numbers is always odd. Starting from 9 + 10, the students tried two more cases, 1 + 2 and 4 + 5, and found that the sum in each case was odd. They went on to try one more case, 11 + 12, as evidence that the conjecture was true. The argument ended with these four cases. When students start to use cases more systematically, they are making Level 2 justifications, which we discuss next.

FIGURE 6.6 Randomly Generated Cases

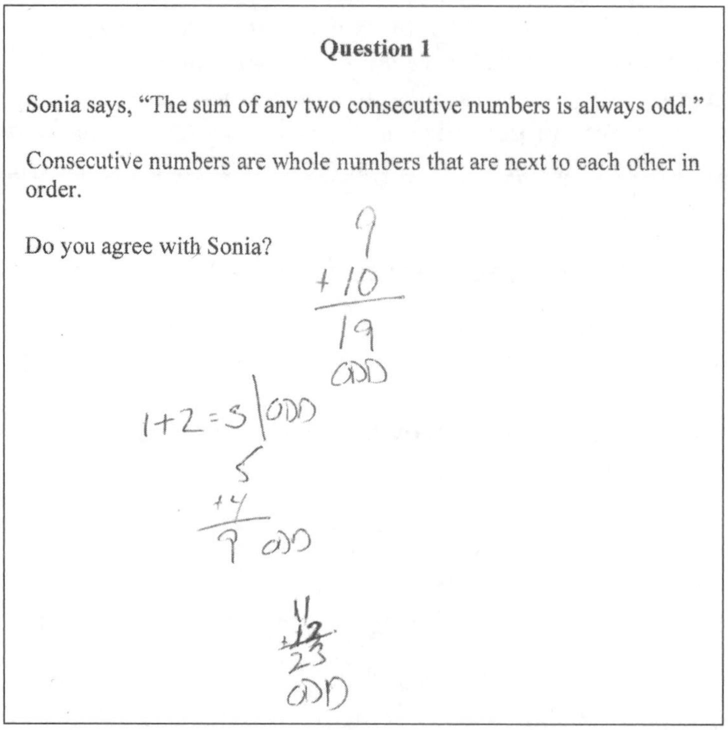

Question 1

Sonia says, "The sum of any two consecutive numbers is always odd."

Consecutive numbers are whole numbers that are next to each other in order.

Do you agree with Sonia?

$$\begin{array}{r} 9 \\ + 10 \\ \hline 19 \\ ODD \end{array}$$

$$1 + 2 = 3 \mid ODD$$

$$\begin{array}{r} 5 \\ + 4 \\ \hline 9 \quad ODD \end{array}$$

$$\begin{array}{r} 11 \\ + 12 \\ \hline 23 \\ ODD \end{array}$$

LEVEL 2: PARTIALLY GENERALIZED JUSTIFICATIONS BASED ON CASES

Level 2 justifications still rely on cases, but students use them to get insights to create partially generalized arguments. This level is critically important for middle schoolers, because it builds on their inclination to use cases. There are three kinds of justifications at this level.

Using Extreme Cases to Push for Generalization

As students start to make Level 2 justifications, they may use extreme cases to test out boundaries of conjectures (Lockwood, Ellis, Dogan, Williams, & Knuth, 2012). For example, to test out a conjecture about the sum of the measures of the interior angles of a triangle, a student might try two angles that are each close to 90° and see that the sides of the triangle are close to parallel, which would prevent making a triangle at all because no third angle is possible. From this extreme case, students can see that the sum of three interior angles in a triangle can't be larger than 180° because two 90° angles can't make up a triangle.

Extreme cases can be useful in number topics as well. With conjectures about number properties, students might attempt large numbers to see if a conjecture is still true for unlikely cases. For example, given the conjecture

> For positive numbers, if a denominator is larger than a numerator, then the fraction is always less than 1

students might try two numbers that are large and very close to each other (e.g., $\frac{999}{1000}$) where the difference is tiny to see if the conjecture still works. This would lead students to make sense of the fractions by visualizing a very small piece left out of a whole, as the numerator and denominator get larger and larger.

In Figure 6.7, students tried making both an elongated parallelogram and a square, two extreme cases, to measure the line segments that make up the diagonals to check whether a conjecture about the bisectors still would hold true. If students actually cut or fold the parallelograms to find which of the formed triangles are congruent, they will gain insight into why the diagonals bisect.

FIGURE 6.7 Creating Extreme Examples for the Diagonals of a Parallelogram Conjecture

Conjecture: The diagonals of any parallelogram bisect each other. Is the conjecture true?

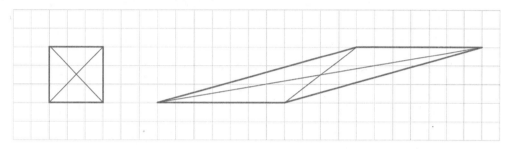

Extreme cases can be the first push toward generality, though the resulting arguments may only hint at generality. Some researchers even categorize them as Level 1, but we consider them Level 2.

Citing Patterns That Reveal Mathematical Structure

Students can use patterns that they see across cases to provide insight into why a conjecture must always be true. In Figure 6.8, a student investigated a conjecture by examining cases where, x is replaced by a number that is less than, equal to,

FIGURE 6.8 Patterns That Reveal Structure

Question 3

Consider the equation $x + 8 = n$, where n is a negative number. What must be true of x so that the equation is true?

Pam conjectures that x must be less than -8. Is she right?

$\times$

$-9 + 8 = n$ $-9 + 8 = -1$
$-10 + 8 = n$
$-11 + 8 = n$
$-12 + 8 = n$
$-13 + 8 = n$

When x is greater than -8, n is positive and gets bigger. When x is -8, n is 0. When x is less than -8, n becomes negative and gets smaller.

or greater than –8. The justification moves beyond simple trials of cases yet isn't fully general because it is missing an explanation about why the properties hold for numbers not tested. In contrast, a fully generalized justification might include solving for *x* and explaining what the resulting algebraic equation shows about *x*.

In Figure 6.9, students chose cases that showed a pattern of sums—that the sums are consecutive odd numbers. There is more regularity to their choices than those of students in the previous section. When interviewed, these students recognized a pattern: that the sum goes up by two each time, resulting in a sequence of odd numbers. Because the pattern is shown in a clearly organized way, and students make a claim based on the pattern, we categorize this as a Level 2 justification.

FIGURE 6.9 Choosing the Right Cases

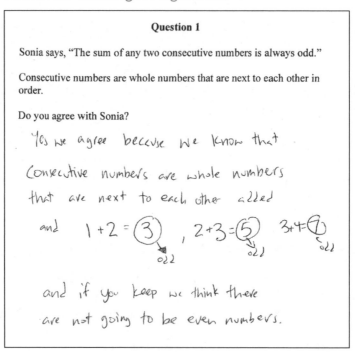

Using a Generic Case to Generalize

Students may focus on one typical example—often called a "generic case"—and examine its structure in a way that reveals key mathematical insights that lead to a more general argument (Balacheff, 1988; Zaslavsky, Aricha-Metzer, Thoms, Sabouri, & Brulhardt, 2016).

Let's say students conjecture that

> Any number that is divisible by 4 is also divisible by 2.

A single case of, say, 20 is not enough to be convincing that the conjecture is *always* true. Couldn't there be examples where the statement is not true? However, students can use 20 as a representative for any number to construct a more general argument, as in Figure 6.10.

> Why does the fact that 20 is divisible by 4 mean that it is also divisible by 2? Because the number 4 is equal to 2 times 2. Since $20 = 4 \cdot 5$, it's also true that $20 = 2 \cdot 2 \cdot 5 = 2 \cdot 10$. We could choose any number that is divisible by 4 and write it out with factors of 2 in the same way.

FIGURE 6.10 Generalizing From One Case

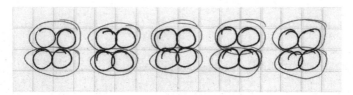

A key insight for this Level 2 justification is that 20 can stand in for any number divisible by 4—there is nothing special about it.

LEVEL 3: FULLY GENERALIZED JUSTIFICATIONS

Level 3 justifications are fully generalized, rely on definitions and properties, and make a logically coherent story. To make these justifications, students draw on statements that have already been accepted as true by their class. They use sound logic to connect what is known to be true to what is yet to be proved.

The classic distinction between types of argumentation is between inductive (based on cases) and deductive (based on logic) arguments. The research we have cited considers how students actually go about making arguments, which blurs the lines between these two—especially at Level 2. But Level 3 arguments are squarely in the deductive category.

While middle school students' Level 3 justifications may not look like the proofs high school students or mathematicians would write, they should share three essential elements with them (Stylianides, 2007). They should

1. Be based on statements and definitions that everyone in the classroom community has already agreed on

2. Use forms of reasoning that are mathematically valid yet comprehensible to students

3. Include representations that communicate to others

Using Definitions

In making Level 3 justifications, students use previously established definitions of mathematical terms to reason about new conjectures. For example, students can justify that all squares are rectangles by saying,

> I think squares are rectangles because they have four sides and four right angles.

This justification relies on a definition of rectangles. Whether this is the definition the class has agreed on would still need to be negotiated.

Using "If . . . Then . . ." to Make Logic Explicit

Part of moving into Level 3 is beginning to use "if . . . then . . ." statements to make explicit the logical connections between statements. Students can use the "if . . . then . . ." structure of a conjecture to find their way into a justification. What follows "if" is the condition that is assumed to be true or that has been

A note about logical language: The following statements—

If a shape is a square, then it is a rectangle.

All squares are rectangles.

A square is also a rectangle.

— are all logically equivalent to adults. We don't recommend the use of the last one with students, because they can fail to see this as a statement about all squares.

established as true. What follows "then" is what students want to show must be also true.

Regarding squares and rectangles, a conjecture in "if . . . then . . ." form would be

If a shape is a square, then it is also a rectangle.

A Level 3 argument for the conjecture is

Let's imagine a shape that is a square. Then we know that it has four equal sides and four right angles. That means it has two pairs of equal sides (the pairs just happen to be equal to each other) and four right angles, which is the definition of a rectangle. So, if it is a square, then it has to be a rectangle.

Notice the statement that begins "Let's imagine." It takes into account what follows "if" in the conjecture. Notice also the prominent role of definitions in the argument, which are more pronounced in Level 3 justifications.

Using Variables

Other Level 3 justifications use variables to represent all cases instead of individual cases. In Chapter 4, in Ms. Cooper's class, students argued that it's possible for the coordinates of a rectangle to include only three different numbers. They represented the coordinates as (a, b), (c, b), (a, d), and (c, d). They then identified a rectangle for which $a = b$. They used variables to represent all possible cases, creating a generalized argument that qualifies as Level 3.

In Figure 6.11, a student uses the variable n to denote *any* number. Also, instead of using two different letters for two numbers, she uses n and $n + 1$ to represent

FIGURE 6.11 Using n to Denote Any Number

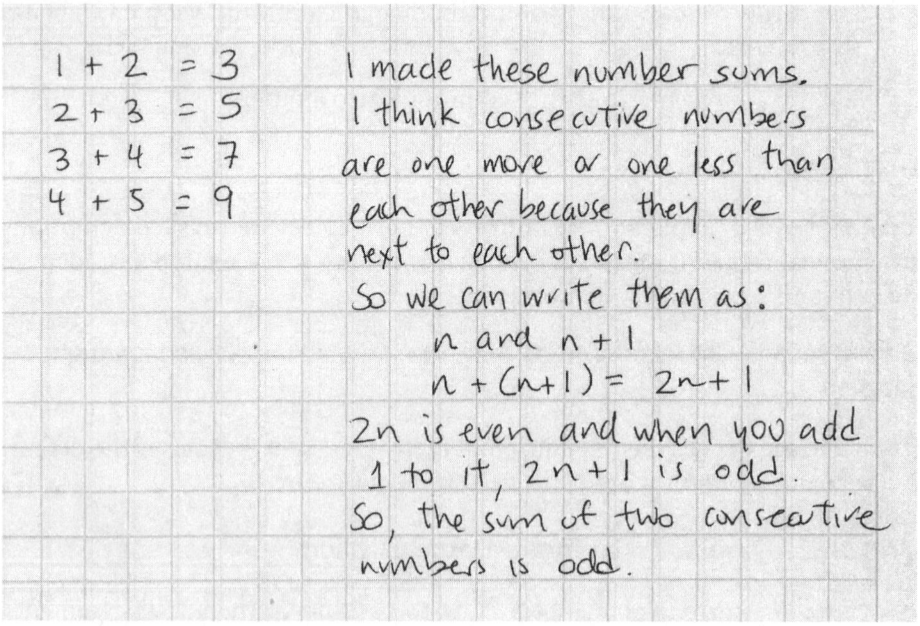

$$1 + 2 = 3$$
$$2 + 3 = 5$$
$$3 + 4 = 7$$
$$4 + 5 = 9$$

I made these number sums.
I think consecutive numbers are one more or one less than each other because they are next to each other.
So we can write them as:
$$n \text{ and } n + 1$$
$$n + (n+1) = 2n + 1$$
$2n$ is even and when you add 1 to it, $2n + 1$ is odd.
So, the sum of two consecutive numbers is odd.

MATHEMATICAL ARGUMENTATION IN MIDDLE SCHOOL

consecutive numbers. Then she applies the algebraic property "adding like terms" to show why the sum must be odd.

By moving beyond the numeric cases and capturing something true about all of them with symbolic variables, the student could then use algebraic manipulations (adding like terms) to form a Level 3 argument.

Going back to the conjecture about a number that is divisible by 4 also being divisible by 2, a Level 3 justification could use a more generalized diagram (representing any such number) or a symbolic representation. Figure 6.12 shows what a generalized diagram could look like. It uses ellipses to indicate that the pattern can be extended indefinitely.

FIGURE 6.12 Divisible by Four

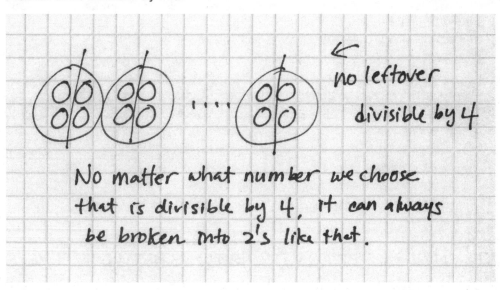

Here's another Level 3 justification, written with algebraic symbols:

> For any number that is divisible by 4, we can write it as $4 \cdot a$, where a is some number. Since $4 \cdot a = 2 \cdot 2 \cdot a$, the number is also divisible by 2.

A RUBRIC FOR LEVELS

Figure 6.13 (see the following page) is a rubric that summarizes the characteristics of a justification at each level. Not all characteristics need to be present to classify an argument at a specific level. Of course, the details we have elaborated need to be considered when classifying a justification, but the table provides initial indicators for use as you are initially sorting a classful of justifications.

FIGURE 6.13 A Rubric for Levels

Level 0: No justification

Shows evidence of

- *repeating the conjecture or providing a new conjecture*
- *relying on external authority—teacher or textbook only*
- *using visual evidence alone*

Level 1: Case-based justifications

Shows evidence of

- *citing more cases only*
- *randomly choosing cases*

Level 2: Partially generalized justifications based on cases

Shows evidence of

- *using extreme cases to push for generalization*
- *explaining patterns across cases that reveal mathematical structure*
- *using a generic case to generalize*

Level 3: Fully generalized justifications

Shows evidence of

- *citing previously established statements and definitions*
- *using language of logic such as (but not only) "if . . . then . . ."*
- *using variables to generalize about a set of cases*
- *using variables, diagrams, or other representations to generalize about a set of cases*

TEACHING MOVES FOR TRANSITIONS BETWEEN LEVELS

What follows are teaching moves to help students make more sophisticated justifications, moving them from one level to the next. You will recognize some that we introduced earlier in the chapters on justifying.

TRANSITIONING FROM LEVEL 0 TO 1

To help students' justifications move out of Level 0, help them see the need for *mathematical* justifications of conjectures. Setting the norms that we discussed for justifying in Chapter 4 is important.

Teaching moves to advance justifications beyond Level 0 include

- Using the basic "why" questions to motivate the need for justification, and using them repeatedly until students start to provide justifications on their own
- Reminding students that in your class, it's not good enough to cite the teacher or the textbook because those are not mathematical reasons that a statement is true or false
- Using the saying, "Convince yourself, convince your friend, convince a skeptic" (Mason, Burton, & Stacey, 1982, p. 8), which helps make clear that justifications are made for a community of mathematical thinkers

MATHEMATICAL ARGUMENTATION IN MIDDLE SCHOOL

- Pushing beyond simply relying on a picture by asking
  - "What mathematical features are the same about the diagrams—or different?"
  - "What would you do to draw a new shape for which the conjecture is also true?"

TRANSITIONING FROM LEVEL 1 TO 2

Because so many of your students probably will be making justifications at Level 1, it's critical for you to recognize arguments based on cases and encourage students to organize their cases and create a more generalized argument. The section on using cases in Chapter 4 provides some moves and we include more here:

Challenging students to go beyond a few cases by asking,
- "Would it be true for other cases that you haven't tried?"
- "Why does it make sense that it works for these cases?"
- "Is there a way you can explain it so that you don't have to try any more cases?"

Asking about extreme cases
- For some conjectures about numbers, you can ask,
  - "Does it work for zero?"
  - "How about negative numbers?"
- For some conjectures about geometry, you can ask,
  - "Is it true for regular polygons as well as polygons with one very small angle?"

Helping students use cases more systematically by
- Asking students to differentiate between typical cases and special cases
- Asking students to look for extreme cases. Extreme cases are a form of special cases. For example, you might ask,
  - "What is a very small/large number for which the conjecture is true?"
  - "What is a very small/large angle for which the conjecture is true?"

When they are stuck making arguments, encourage students to pick a simpler version of the problem, then argue that, for example, a problem with big numbers, or a problem using simple numbers and argue with those.

—Seventh-grade teacher

Drawing attention to patterns by asking,
- "What do all these have in common?"
- "What stays the same as _____ changes?"
- "What if you increase the numbers by (for example) 2 each time; can you see a pattern?"
  - "What if you increase the angle by (for example) 10° each time; can you see a pattern?"

Asking students how they know that a pattern will continue:
- "Will the pattern continue? Why is that?"
- "Why does it make sense that the pattern will continue?"

(Continued)

(Continued)

TRANSITIONING FROM LEVEL 2 TO 3

As students become more comfortable making partially generalized arguments that are grounded in cases, you can push them to make arguments that are even more general. As they become more familiar with using variables, they acquire tools for making more generalized arguments.

The section on using logic in Chapter 3 provides moves you can use in conjecturing in support of Level 3 justification and we include more here:

- Prompting students to formulate conjectures in "if . . . then . . ." form
- Explaining that
  - "If . . ." is what you assume or know is true
  - "Then . . ." is what you want to show must always be true
- Asking students what else must be true based on the "if . . ." part of the conjecture, until they get to the "then . . ." part
- Encouraging "if . . . then . . ." statements as part of the chain of statements that leads to the conclusion
- Suggesting the use of variables to express generality: "Can you use variables to say how it works every time?"
- Encouraging students to use definitions: "What is the definition of ___? How can that help us?"

WORKING TOGETHER

In this meeting with your grade-level team, professional learning group, or mathematics department, you will have the opportunity to examine samples of students' work, including work from your own students. Sharing your students' work with your colleagues will allow you to get important comments and their reflections for improving your classroom argumentation practices.

Before the meeting: With your students, use one of the tasks on justifying conjectures presented in Chapter 4. Be sure to make copies or upload the files of one or two samples of your students' work and a copy of the task. Decide which level each sample represents and be prepared to give reasons for your decision based on this chapter.

Exploration and discussion (45 min)

1. In your grade-level team or small group, share one of your student work samples. Explain why you assigned it a level, and discuss with your colleagues. (15 min)

2. Use another work sample to present to the other group members:
   - The presenting teacher describes the student work, without making judgments about the quality of the work or interpreting what the student did. (5 min)
   - The participants put themselves in the role of teachers and try to make sense of what the student was doing and why. Each group discusses the following questions:
     - How did the student interpret justifying?
     - At which level was the student's response?
     - What did the student understand?
     - What did the student have difficulty understanding? (15 min)

3. Based on the group's observations of the student's work, discuss
   - Moves the presenting teacher might use to help the student make a more sophisticated justification
   - Ideas for revising the task
   - Follow-up lessons or tasks the teacher should consider (10 min)

Wrap-up and assignment (15 min)

4. As a whole group, reflect on the changes you plan to make in your own classroom practices: What will you do differently as a result of analyzing samples of students' justifications? Try out some of your ideas before the next session.

NOTES

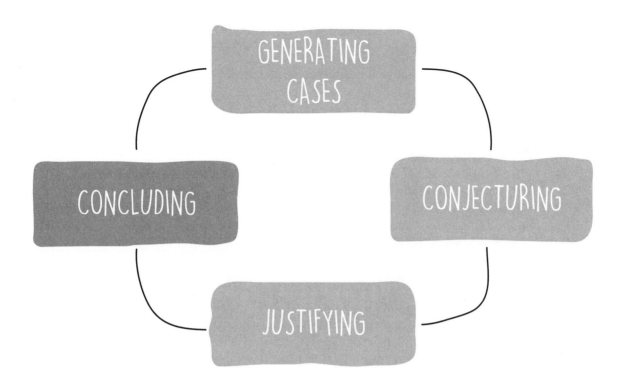

CHAPTER 7

Concluding

In this chapter, you will learn

- What concluding means in the mathematics classroom
- Through vignettes, how students make concluding statements
- Moves for establishing agreement and summarizing the argument
- Norms for concluding, along with games to help establish them
- How to plan for concluding in your lessons
- Tasks for engaging students in concluding an argument
- Through your work with colleagues, additional moves for engaging students in summarizing and concluding arguments

WHAT DOES IT MEAN TO CONCLUDE?

So your class has generated cases and then made a conjecture and then justified it—after all that, they still need to conclude the argument. Concluding is critical, though it is at risk for being given little attention by the time the rest of the work in argumentation is done. In this chapter, we contrast two classroom routines for concluding.

Concluding means deciding whether a conjecture is true or false based on a justification. When concluding an argument, students should summarize the argument—the conjecture and the justification—so all can agree on whether the justification is sound. By sound, we mean that the logical connections between statements make sense, as expressed appropriately for middle schoolers, and that the mathematical statements in the argument are all true. Concluding may also involve taking additional steps: If we know this conjecture is true or false, what is the next related conjecture to explore? This and related questions are discussed in this chapter.

Concluding provides an opportunity to make mathematical connections. You can explain or help students express what an argument reveals about a mathematical concept. You can use an argument as a springboard into new content. You can also help students understand the qualities of good arguments, as exemplified in the argument they've just made. We provide moves for making these connections in the chapter.

VIGNETTES: CONCLUDING

Concluding follows right on the heels of justifying. Rather than present a new vignette on concluding, we review how concluding was done in the vignettes found in Chapter 4 on justifying.

Concluding in Ms. Cooper's Lesson

Recall that Ms. Cooper guided students in justifying several arguments about coordinate geometry, together as a whole class.

> Ms. Cooper frequently prompted students to signal their agreement or disagreement with a thumb up or down.
>
> Ms. Cooper: Does this argument convince you? Thumbs up or down.
>
> All thumbs pointing up signaled the beginning of concluding.
>
> Ms. Cooper helped a student summarize the argument other students had made, writing on a common display:

MATHEMATICAL ARGUMENTATION

Conjecture [. . . Always . . .]

> For the vertices of vertical sides of the rectangles, there were two coordinates the same.

Justification [How do we know?]

> The vertices of a vertical side of the rectangle are always the same distance from the y-axis.
> This is because the vertical side is parallel to the y-axis.
> The x-coordinates tell you the distance from the y-axis.
> Because of that, the x-coordinates are the same for all the points on the side.
> So the x-coordinates of the vertices on a vertical side have to be the same.

Conclusion [True or False]

> True

She called attention to the fact that the argument had convinced everyone and so declared the conjecture to be true.

> Ms. Cooper: We said this justification convinced us all, so we can fill in the conclusion.

She instructed students on qualities of a good justification.

> Ms. Cooper: Notice this justification started with something we knew for sure and ended with the conjecture we wanted to establish as true. Every step followed from the one before it. That's what a good justification that a conjecture is true should do—you should be convinced every step of the way.

Concluding in Mr. Flores's Lesson

Recall that Mr. Flores had students, in a whole-class discussion, present and critique the arguments they had made in small groups for why the distributive property works.

Mr. Flores declared an argument as concluded when there were no remaining questions for the arguers. He had established the norm that students should keep asking questions until they were convinced—that is, until they had no remaining doubts about the soundness of the argument.

> **Mr. Flores:** Does anyone else have questions for Rayhan and Sara? No? Then their argument based on a diagram was convincing and they showed the conjecture is true.

As an extension to concluding, Mr. Flores had students compare two arguments for the same conjecture after the class had accepted them both as true.

> **Mr. Flores:** Can you explain how the two justifications show the same thing, even though one is a diagram and one is in symbols?
>
> **Cesar:** They both use doubling the amount in the parentheses.

Mr. Flores had students say how doubling was represented in each justification. Then he helped students further generalize. Finally, he had them apply their generalization to use the distributive property on two more algebraic expressions, providing an opportunity for meaningful practice.

> **Mr. Flores:** Cesar is saying that we could replace 2 in the expression with any number. So really, we have established something more general than just the original conjecture. Now, use this idea to write two other equations that show how the distributive property works with variables.

TEACHING MOVES

Moves for concluding include methods for establishing agreement, eliciting summaries of arguments, and helping students formulate their next conjectures based on their conclusions.

ACCESS FOR ALL

Broadening Access Through Concluding

Both Ms. Cooper and Mr. Flores used concluding as an opportunity to broaden access to mathematical authority. Ms. Cooper relied on students' gestures to indicate agreement, in this way, including students who may have been reluctant to speak up. She honored students' status as a mathematical community by introducing standards from the broader community of mathematicians in her comments on what made a good justification. Mr. Flores pressed students to voice disagreement, setting a norm that disagreement with ideas, not people, was an important part of argumentation. He leveraged students' argumentation to lead them to a more advanced view of the distributive property.

Establishing Agreement

It can be tricky to establish that everyone is in agreement about the truth status of a conjecture—whether it is true or false.

What if not every student is in agreement with a conclusion? Despite the presentation of fairly convincing arguments, sometimes students may disagree with them. We have seen teachers do some things we wouldn't recommend, such as having the class vote on the truth of a conjecture. This isn't the way mathematicians decide on truth. Instead, you can try asking the "hold-out" students to pinpoint exactly where in the presented justification they find their disagreement and further the class's argumentation to a conclusion agreed upon by all.

What if you, as the teacher, don't agree with what the students agree on? It's time then for you to give the class the benefit of what you know from your engagement with larger mathematics communities (for example, what you learn from reading this book). You should go beyond simply saying students are wrong. You should present an argument for why your conclusion is different from that of the class or elicit such an argument from the class once you have presented the alternative conclusion.

What if everyone seems to agree on a justification for a nonmathematical reason? Students may be convinced by a faulty justification put forth by a student who is perceived to be "good at math." You may need to offer a counterexample yourself to help students relaunch into arguing about the conjecture independent of who is making the argument. You may also find that some students, because of their own perception of their ability as math learners or because of shyness, are not willing to publicly disagree with the class consensus (Civil & Planas, 2004). In that case, some tried-and-true methods such as "think-pair-share" may provide opportunity for more students to have a voice. You can also cite a student's argument anonymously: "I heard someone make this argument today . . ."

Sometimes, students may have argued from an idiosyncratic definition of a mathematical term. You could introduce the more widely accepted definition and explain how this changes the argument. For example, as we have discussed, students may define rectangles as quadrilaterals with two different side lengths and, using this definition, conclude that it is *not* true that all squares are also rectangles. Students should know what definitions are considered standard by the larger community of mathematicians. Their examination of how a different definition leads to a different conclusion can help solidify the role of definition in argumentation.

Here's another example: You wouldn't want to let the conclusion stand that the sum of the measures of the interior angles in a triangle is 179°, which students may conclude if they are basing their argument solely on the measurements of a few triangles they examined. By introducing the correction that mathematicians have concluded that the sum is 180°, and asking what a 180° angle looks like, you can suggest a justification that goes beyond a couple of examples.

Moves for establishing agreement include

- Creating a way for students to express their agreement, such as thumbs up or no remaining questions
- Asking "hold-out" students to say as precisely as they can where their disagreement lies, and asking other students to address these points of disagreement
- Providing students with corrective insights and new arguments so that they don't leave class with some incorrect mathematical information: "We've decided that ___ is true, but most mathematicians don't agree and here is why." (Or, ". . . can you see why?")

Summarizing the Argument

A good conclusion includes a summary of the class's argument that is understandable to everyone in the class. A summary consists of the conjecture together with its justification and its truth status. This is the time to make clear the logical connections between steps in the argument. Justifications emerge from classroom discussion in what Lakatos (1976) calls a "zigzag" fashion—ideas may pop up and later reemerge. But the summary should help students see the connections from one idea to the next in a more orderly fashion.

In the vignette, Ms. Cooper's class's summary started with what students knew for sure about vertical lines and proceeded to the conclusion that the vertices of a vertical side have the same x-coordinate. Everything they had discussed was put together in a logical progression: stating that the vertices on any vertical side are the same distance from the y-axis because the side is parallel to the y-axis and relating this statement to the x-coordinates of the vertices.

Consider another example, for an argument about linear functions:

> We all agree that the conjecture—the steeper line represents faster speed—is true. This is because each point on the steeper line shows a greater distance than on the less steep line, for the same time traveled. And when you cover a greater distance in the same amount of time, you are going faster.

Notice that the connecting words *because* and *when* establish logical connections between statements. This is the level of logical connection that is appropriate for middle school. Notice also that this summary begins with the truth status of the conjecture and leads back to what students already know, in contrast to Ms. Cooper's class's summary, which began with what students already knew and ended with the conjecture. Both kinds of summaries work.

To make clear the truth status of a conjecture, we have promoted the idea of "stamping" a conjecture as true or false—for example, with a sticky note labeled T or F attached to a poster of an argument. Then, once a conjecture is established as true, students can use that mathematical statement in other arguments. You can then call these once-conjectures simply "true statements" or "conclusions." In later years, students will learn to call these "theorems," though we don't find that term necessary for middle schoolers. In the planning section, we discuss ways for the class to keep track of statements that have been justified as true.

Summarizing an argument includes

- Using "and because of that we know . . . ?" to elicit the argument in order from what students already know (and have established as true) to the conjecture under consideration
- Helping students connect statements logically with words such as "because" and "when"
- Having students who have started using "If . . . then . . ." start the summary with what comes after "if" and end it with what comes after "then"
- Establishing the truth or falsity of a conjecture by "stamping" it as one or the other and, if appropriate, adding it to the list of statements proved true

Explaining Qualities of Good Arguments

Explaining qualities of good arguments is another move to use in concluding. Ms. Cooper commented on the qualities of a good argument that had been demonstrated in the arguments that day. Your commentaries may include some of the following:

- "We justified this conjecture by using the definition of _____."
- "We showed this conjecture is false by using a counterexample."
- "We learned that we can't simply use a couple of cases to prove a conjecture."

Leading to Another Conjecture

Sometimes a new conjecture to consider is immediately obvious based on one that students have justified. In the vignette, Ms. Cooper's class's conclusion about vertical lines led naturally to consideration of horizontal lines, which then led to consideration of both types of lines in a rectangle. Ms. Cooper took advantage of that progression of conjectures, laying them out for students in order.

In another example, say students have argued that the graph of an equation of the form $y = kx$ must be a straight line through the origin. This could lead to conjectures about the nature of equations for straight lines that do not go through the origin. In this way, students can connect proportional and linear functions.

Some moves for generating new conjectures based on conclusions include

- When a conjecture turns out to be false: "Now that we know this conjecture is false, let's consider a related conjecture that we think could be true."
- When a conjecture is proved true: "What does this conclusion make you wonder about next?"

ACCESS FOR ALL

In an earlier chapter, we discussed how to organize students' written records of cases they have generated and their conjectures. You may find that much of justification is done in whole-class or small-group discussion, without students creating much of a written record. That's OK because as you conclude, you can ask students to write summaries of their arguments. This is a time you can address the needs of special groups of students: for example, students who have trouble with organization or ELL students. Of course, helping students keep written records of argumentation is important for all students. Early on, this may be a matter of copying what the class came to as a summary from a public display. Next, you may provide sentence starters. As they become more confident arguers, you can have students write their own summaries individually. You can use these write-ups as exit tasks, which require a higher level of thinking than procedural exit tickets (Fennell, Kobett, & Wray, 2017).

Helping students make written records of their arguments can include

- Having students record their arguments on paper versions of the blank argumentation poster
- Providing sentence starters such as "Because of that . . ."
- Using written arguments as exit tasks

> With concluding, I assign the students to do more sentence writing. For the concluding statement, I will say: tie in what you are learning with the math content. I use this assignment as an exit task.
>
> —Sixth-grade teacher

Making Connections to Concepts and Skills

To make argumentation really work in your mathematics curriculum, you can build on concluding to create opportunities for students to develop concepts or practice skills. For example, as in the task from Chapter 4, when students argue that steeper lines represent higher unit rates, they are building a foundation for the concept of slope, though they won't yet use that word. You can help students develop the formula for slope based on the graphs they have examined as cases, cementing the connection between slope and unit rate.

It is important to help students see the connections between the mathematical representations they are using as they work on conclusions (Inglis, Mejia-Ramon, & Simpson, 2007). In the vignette in Chapter 4, students were working on justifying the conjecture, "For the vertices of vertical sides of the rectangle, there are two coordinates the same." First, students gave examples of specific cases in the rectangles they saw on the display. Ms. Cooper then guided them to move from visual representations to the properties of vertical lines. By doing this, they were able to conclude that the x-coordinates of a vertical side would always be the same. The students did not abandon their argument based on visually observed cases, but added to it, using the properties of vertical lines, connected to the definition of parallel lines, to justify the conjecture.

An argument can also lead to meaningful practice. Mr. Flores had students write two more examples of the distributive property using factors other than 2. Their justifications helped them apply the property correctly, knowing that the number outside of the parentheses had to apply to both numbers inside the parentheses in expressions of the form $a(bx + c)$, although they did not explicitly use the constants a, b, and c.

To connect an argument to a concept or skill, you can

- Assign related practice problems in which students use elements of the justification they just made
- Ask, "What do we know about _____ now that we have proven this conjecture is true?"

ESTABLISHING NORMS

We recommend that you establish the following norms and routines for concluding, using the following warm-up games.

Norms

- ★ **Know when to stop.** Students may know when to stop an argument in nonmathematical situations, such as storytelling. With mathematical arguments, the argument ends when the conjecture is established as true, and everyone agrees on this, and students should start expecting this.
- ★ **Retell the argument from beginning to end.** The process of justifying a conjecture is sometimes messy, especially when negotiating many student voices. So, retelling the argument is an important part of concluding. The retelling—or summary—includes the conjecture, the justification in a logical order, and the decision about the truthfulness of the conjecture.
- ★ **Base your conclusions on what is said, not who said it.** Retelling the argument without giving attributions to individuals helps students focus on the mathematical truth of the statements rather than the status of the student who made the statement.

Warm-Up Games

Warm-up games based on storytelling can help students solidify the idea that a mathematical argument should have a beginning, middle, and end. Here are two storytelling games that support the norms for concluding and connect summarizing or retelling an argument to telling a story.

In Word at a Time (e.g., Johnstone, 2012), students see a concrete example of stopping an argument when everyone agrees.

WORD AT A TIME

Players stand in a circle. The goal is to make a sentence collectively.

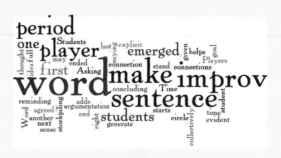

- The first player starts the sentence by saying one word.
- The next player adds another word that would make sense given the first word, and players continue adding words until a sentence has emerged.
- When the end of the sentence is evident, the last player says, "Period."

Asking students why they thought the sentence had ended, or why one student said period, helps make connections to concluding. Students may say that there was no more to say or that a full idea had emerged. You can make the connection to argumentation explicit by reminding students about how they agreed that "period" was said at the right time.

In another game, String of Pearls (e.g., Hall, 2014), students see how a story can be fleshed out from the starting and end points, very similar to the way an argument may unfold. At the end of the game, students tell the whole story from beginning to end, regardless of the order in which the story was created.

STRING OF PEARLS

In this game, a story is told in five sentences, each added one at a time as people fill in the spots in a lineup.

- The first player is asked to give the first sentence of a story that has never been told and stand on one end of an imaginary line on the floor.
- A second player stands at the other end of the line and says the final sentence of a story. (It need not appear to be connected to the beginning.)

- A third player stands somewhere between them on the line and gives a middle sentence of the story that could connect the first two sentences.
- Then, the three people say their sentences in order down the line. The story now has three sentences.
- Next, a fourth and a fifth player step into the line to tell the parts of the story that are missing between the beginning, middle, and end.
- At the end, a single, unified story is told again with each player saying his or her own sentence.

Students can see how a story can get more complex as they fill in the gaps. There is a very simple, blunt story when they tell only the beginning and the end. The story makes more sense as they fill in the middle and the connections that need to be made become more obvious. This is similar to how their arguments may emerge in class: Different students will contribute parts of the argument, and sometimes it's not quite clear if the contribution is the beginning, the middle, or the end until someone else makes the second contribution. In the end, though, the story is read out in order, just like concluding an argument is done.

PLANNING

Concluding can easily be the neglected stage in argumentation. When getting to the end of an argument, it may feel sufficient to just label the conjecture as true or false and move on. But by doing that, you miss all the opportunities just discussed, including connecting to content. You can plan for those opportunities. In preparing for helping students to conclude their arguments, you'll want to choose and establish a method for concluding, consider connections to new conjectures and content, and plan for how to help students make written conclusions and keep them organized.

Establishing a Method for Concluding

You saw the two different ways that Ms. Cooper and Mr. Flores established agreement. Plan to use either the "thumbs-up/-down" method, or the "no further questions" method—or one of your own choosing—consistently, so that students will know what to expect. Plan for how to introduce and establish your chosen routine. You also need a basic method for when disagreement remains. Build into your lesson opportunities to help students learn to respectfully disagree until they are convinced by an argument. Plan for the use of sentence starters such as, "I disagree with [statement, the argument] because . . ." (notice that the disagreement is with a statement, not with a person). Put these sentence starters on a public display before class begins.

Connecting to Content

In preparing to teach a series of argumentation lessons, you will have already planned for your target conjectures and the possible justifications of them. What can your students do once they have concluded those arguments? These are all important questions to consider:

- Will conclusions relate closely to content coming up in your curriculum? For example, the argument above about steeper lines relates to unit rate and slope.
- Do the justifications rely on any new reasoning you want to point out to students? For example, if students are beginning to use "If . . . then . . . ," you can reinforce its use in upcoming lessons.
- What new conjectures could students make based on those they have concluded the truth or falsity of? For example, as discussed, students could investigate properties of linear functions once they have established properties of proportional relationships.

Recording Results of Conclusions

You should plan to have a designated place for statements that have been established as true through the process of concluding. This may be a public display, or you can have students keep a list of statements that have been established as true in their notebooks—or even better, both. Or, using a shared digital document, students can collaboratively keep the list of statements the class has proven true. This is a great way to help students develop their own sense of mathematical authority, as they gain confidence in adding statements to the shared document. This list can then be accessed from devices used by small groups or individuals, as well as displayed for the whole class to use collectively.

This record keeping is important because all statements that have been established as true can be used in future arguments. You'll want to designate a specific time when statements can be added to the list, such as after they have been stamped true or false or during the last 5 minutes of class.

Delaying a Conclusion

Sometimes, you will get to the end of a class period without concluding an argument. Be prepared for this by planning for different options. Will you tell students the conclusion rather than leave them with a potential misconception? Or, will you make clear the uncertain truth status of the conjecture at hand, preparing students to finish the argument the next day? Each can be productive, depending on where you are in your curriculum. Sometimes, you simply can't spend any more time on an argument, and you will want to summarize the end of it for students. If you leave enough time to do this, students can draw their own conclusion from your summarized argument. Other times, important insights are to be gained from a more elaborate articulation of an argument, and it will be worth leaving it for the next day. Planning in advance for possible justifications of your target conjectures, as discussed in Chapter 4, will help you know how to choose a path in the moment.

TASKS

Any of the tasks from the other chapters will entail concluding, after students engage in the justifying stage. Another way to engage students in concluding is to present ready-made arguments and ask them to decide whether or not they support a particular conclusion—in other words, to critique the arguments, as it says in the standards. We present some of these ready-made arguments that will provide opportunities to advance students' conceptual understanding of content.

TASK 7.1 FOR THE DISTRIBUTIVE PROPERTY USING VARIABLES, AN INCORRECT ARGUMENT FOR STUDENTS TO CRITIQUE

Critique the following argument:

Conjecture: For all values of x, $2(3x - 1) = 2 \bullet 2x = 4x$ is always true.

Justification: This is because $3x - 1 = 2x$, and so then $2(2x) = 4x$.

This conjecture is false, and the justification is flawed because $3x - 1$ will not always equal $2x$. Encourage students to find a numerical counterexample by finding a value of x where a statement in the argument is false. For example, when $x = 5$, $3x - 1$ is $15 - 1 = 14$ and $2x$ is 10. This argument points to the importance of using the words "for all values of x" in front of the conjectured false identity, $2(3x - 1) = 4x$, because that helps trigger using a counterexample as a strategy. Then students can try an alternative conjecture and justification as described in Chapter 4.

TASK 7.2 FOR COMPUTING WITH SIGNED NUMBERS, AN INCORRECT ARGUMENT ABOUT A PARTICULAR CALCULATION FOR STUDENTS TO CRITIQUE

Critique the following argument:

Conjecture: $3 - 4 = 1$

Justification: This is true because $3 - 4$ is the same as $4 - 3$.

This argument depends incorrectly on the reasoning that subtraction is commutative and on using the elementary school reasoning that you should always subtract the smaller from the larger number.

You can guide students to correct reasoning by asking whether $3 - 4$ and $4 - 3$ *should* give the same result. Students will be confident that $4 - 3$ is 1. They can show why using a number line. Then, they can use the number line to consider how $3 - 4$ would give a different result, mirroring the actions on the number line for $4 - 3$. Or students could consider a real-world context such as money.

TASK 7.3 FOR TRANSFORMATIONS, AN INCORRECT ARGUMENT FOR STUDENTS TO CRITIQUE

Critique the following argument:

Conjecture: When you reflect any point with positive coordinates over the line $x = 3$, the resulting image has two positive coordinates.

Justification: I know this because I tried reflecting the point (1, 2) over the line $x = 3$, and the image is the point (5, 2).

This argument is made based on one example, which is never sufficient for proving a generalized conjecture to be true. In Chapter 6, on levels of argumentation, we said much more about justifications based on examples only. You can use this argument to help students see the limitations of arguing from one example. With a diagram or dynamic geometry representations, they can look at the whole plane and see that the image of many points will have negative coordinates.

WORKING TOGETHER

Working with your professional learning community, department, or grade-level team, think about moves for summarizing and concluding for the grade levels you teach. Be prepared to share your ideas with your colleagues.

Begin by selecting a conjecture from a vignette from a previous chapter or from the list of arguments in this chapter that reflects common early conceptions students may hold.

Exploration and discussion (45 min)

1. With a partner, justify the conjecture. After you make your justification, make a written summary including the justification and the truth status of the conjecture. Then evaluate your summary:

   - Do the logical connections between statements make sense?
   - Are the mathematical statements accurate?
   - Does the conjecture need to be revised? If so, briefly explain.
   - If not, what is the next related conjecture that you might explore? (20 min)

2. Review the moves for concluding that are used in the vignettes. Then design your own moves, questions, or prompts to elicit conclusions for your justification. (15 min)

3. How will you use the classroom norms in this chapter to help students in summarizing and concluding their arguments? (10 min)

Wrap-up and assignment (15 min)

4. As a group, discuss some of the moves that you will use to help your students

   - Make a transition between justifying and concluding statements
   - Explain how the mathematical representations they used helped them in justifying their conjectures (10 min)

5. Select a task or ready-made argument that you will assign over the next few days to engage your students in concluding an argument. (5 min)

NOTES

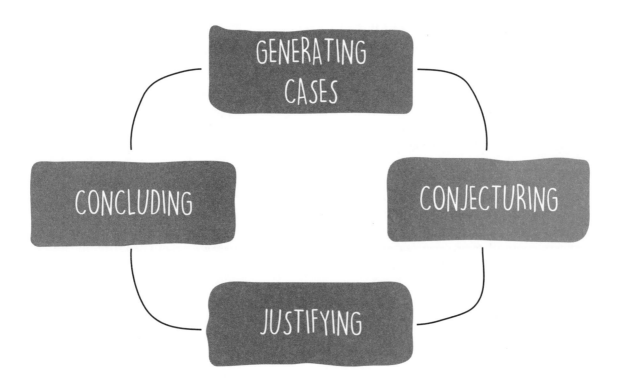

CHAPTER 8

Planning

In this chapter, you will learn

- Steps to use in planning argumentation lessons and argumentation *in* lessons
- How to use written plans for your mathematical argumentation lessons
- Through a vignette, techniques for visualizing your lesson and refining written plans
- How to plan your lessons for supporting different groups of students, including struggling students
- An approach for updating your plans for future argumentation lessons and sharing them with colleagues
- How to sharpen your argumentation teaching skills through meetings with your department or grade-level team

HOW CAN YOU PLAN FOR STUDENTS' ARGUMENTATION?

Throughout this book, we have encouraged you to help your students come up with their own mathematical conjectures, justifications, and conclusions. This means, of course, that the unexpected will happen in your mathematics class. That's when you most need your skill as a *disciplined improviser*, as we discussed in Chapter 1. And that's why planning for students' argumentation is so important: Planning is part of the discipline needed to successfully improvise in the classroom. This planning requires more than identifying a good task, although that is important, and we have provided many tasks ready for your use. It requires thinking about the teaching moves you will need to help students through the stages of argumentation, possible arguments students will make, and how to respond to students' arguments. You will also need a ready supply of teaching moves to respond to unanticipated conjectures and justifications. In the previous chapters, we've provided a wealth of teaching moves and ways to incorporate them into your teaching. But you can't hold all of that in your head as you stand before the class. Advance planning helps you respond when that student in the back of the room says the most amazing thing—or, on the other hand, when no one has anything to say.

When you *visualize* a lesson, you imagine what will happen in the classroom, as it might actually happen.

This chapter focuses on two fundamental planning practices that work together to help you successfully teach for argumentation. *Written plans* enable you to write down the mathematical territory you want to cover, anticipate student arguments, and plan for some basic teaching moves. Then, *planning aloud* with other teachers allows you to imagine a path through a lesson, bringing in teaching moves in ways that act as a rehearsal for the real thing. We call this planning aloud "visualizing" because you should actually imagine what will happen in the classroom, as it might unfold.

WRITTEN LESSON PLANS

It's important to write out a complete plan that guides your teaching of a lesson that includes argumentation—or really any lesson. Lesson plans are a tool for you to use in teaching, not simply an exercise in filling out forms. They help you ensure that you are teaching the content and practices/processes that you set out to do. You may be using an existing lesson plan and adapting it to include conjecturing or justifying, or you may be planning based on an argumentation-focused lesson, such as from the Rectangle Coordinates activity featured in the vignettes in this book. We have included a form (available for download at resources.corwin.com/mathargumentation) that you can use for planning your lessons. Figure 8.1 shows it filled in for Ms. Cooper's lesson from Chapter 2 on generating cases.

FIGURE 8.1 Example Lesson Planning Template

Lesson	Generating cases for patterns in vertices of coordinates
Standards	Apply these techniques in the context of solving real-world and mathematical problems. Understand signs of numbers in ordered pairs as indicating locations in quadrants of the coordinate plane.
Resources	Handout—Rectangle Coordinates
Learning goals	Draw rectangles parallel to the axes by plotting the vertices and connecting them. Recognize and state patterns found in the coordinates of the vertices of these rectangles, including • x-coordinates are the same for vertices of vertical sides • y-coordinates are the same for vertices of horizontal sides • There can be four different coordinates across the vertices, and they follow a pattern • For rectangles with their center at the origin, the x-coordinates are "opposites" and so are the y-coordinates
Related prior knowledge	• Graphing in Quadrant I • Terms *axes* and *origin* • Recognize what a rectangle is, though maybe not a formal definition • Recognizing parallel and perpendicular lines
Vocabulary to be introduced/ reviewed in the lesson	• Quadrant • Vertex/vertices • x-coordinate; y-coordinate
Stage of argumentation	Generating cases

Warm-up game: Off the Shelf	Time: 5 minutes Grouping: Whole class
Teaching move	**What students may do and what I will do in response**
Play the Off the Shelf improv game. Ask students what this game has to do with doing mathematical argumentation.	Students may say that they have to be creative. I may need to remind them that other school norms apply when playing this game.

Introduction: Introduce four-quadrant graphing	Time: 7 minutes Grouping: Whole class
Teaching move	**What students may do and what I will do in response**
Review coordinate graphing and include negative numbers on axes.	Draw axes and label them. *I will need to think more about where students can go wrong here.*
Have all students graph a rectangle with vertices in at least two different quadrants.	Position vertices on either side of one of the axes.
Set up overall activity: Do you think there will be any patterns you can see when we graph several rectangles? How can we know if those patterns always work?	Students may not have many ideas at this point. That's OK . . . move them on to next stage.

Task 1: Students graph four rectangles and label coordinates	Time: 20 minutes Grouping: Small groups
Teaching move	**What students may do and what I will do in response**
Remind students what generating cases means.	Students listen.

(Continued)

Teaching move	What students may do and what I will do in response
Have students read Prompt 1 on the handout.	Students read aloud. Some students may have questions.
Direct them to graph four different rectangles, encouraging them to choose a variety of different rectangles.	Students may not know how to choose a variety of rectangles, but just get them started on one for now.
Visit each group to guide their generation of cases (four rectangles).	Students may focus on special cases; I can have them broaden or simply focus on those special cases and patterns they see. Students may misunderstand the instructions; graph an example rectangle with them. Students may be making rectangles in first quadrant only. Encourage them to choose vertices with negative coordinates as well. Students may be looking for patterns already. Encourage them to see if their pattern holds for an additional rectangle.
Encourage each group to graph a rectangle symmetric about one axis, to ensure coverage of this standard. Ask what is special about its coordinates, as compared with other rectangles they have graphed.	Students may need to be reminded what symmetric means.

Task 2: Students look for patterns in coordinates of vertices across the rectangles	Time: 10 minutes Grouping: Small groups

Teaching move	What students may do and what I will do in response
Ask what is the same for all the rectangles.	(I don't yet know what students may say here.)
Ask if the pattern works for another rectangle (setting them up for a conjecture about all rectangles).	Students may test out the pattern on a "easy rectangle"; ask them to choose a more challenging case with negative coordinates.

Wrap-up	Time: 5 minutes Grouping: Whole class

Teaching move	What students may do and what I will do in response
Ask students to share some of the patterns they saw.	Students may not use all correct vocabulary; revoice with correct vocabulary from above. Students may not state the patterns very clearly. Don't worry too much, we can work on it tomorrow in conjecturing.

 Lesson planning template available for download at **resources.corwin.com/mathargumentation**

There are several important features to the plan, including both information that applies to the lesson as a whole and the concrete steps you will take over the course of the lesson. Let's break it down.

Standards

Think through which parts of which standards you will be addressing. A lesson may address both algebra and geometry standards, for example, so you may want to look closely at those sets of standards for your grade level. You also may not address an entire standard at once. For example, one standard for sixth-grade geometry addresses coordinate geometry:

> Draw polygons in the coordinate plane given coordinates for the vertices; use coordinates to find the length of a side joining points with the

same first coordinate or the same second coordinate. Apply these techniques in the context of solving real-world and mathematical problems.

Ms. Cooper's students would be problem solving by looking for patterns in this lesson. She also noted that she was addressing a standard from number systems regarding negative numbers:

Understand signs of numbers in ordered pairs as indicating locations in quadrants of the coordinate plane; recognize that when two ordered pairs differ only by signs, the locations of the points are related by reflections across one or both axes.

Her lesson plan reflects that she would prompt students to graph a rectangle that is symmetric about an axis, in order to begin to address this standard.

Resources

Your lesson can be based on handouts such as those on the book's website, or it may come from the textbook you use, or yet another source. Indicate it by title or chapter and section here. This will help you reuse the lesson from year to year.

Related Prior Knowledge

In this section, you should put what students already need to understand, know, or be able to do in order to succeed in this lesson. It's important to take time to carefully consider what is really essential for success. This is not a list of everything they already should have learned up to this point, nor should it include the concepts, skills, and knowledge that the lesson will address. That may seem obvious, but it is easy to overreach when writing up this section. Ms. Cooper noted that students needed to be able to make a rectangle, but she did not include knowing the definition of a rectangle, which was fine because the definition would not be required in this lesson.

Vocabulary to Be Introduced in the Lesson

This is a list of new vocabulary that you will introduce over the course of the lesson. It may also include vocabulary that you think students may have forgotten or often get confused about, which you plan to remind them of or clarify with them. Rather than a separate part of the lesson, plan for teaching vocabulary throughout the lesson as you highlight the mathematical nature of students' comments, introduce new words as they are needed, and revoice student statements to be more mathematically precise.

Stage of Argumentation

This may include one or more parts of the four-part model. If you use a premade conjecture, then your lesson will focus on justifying. If you are teaching a lesson in which students create their own conjectures, you may have them justify them as well in the same lesson, and then your plan would indicate you are covering both those parts of the argumentation model. Often, concluding goes along with justifying. If you are teaching a series of lessons that spans the four-part model, then each lesson may focus on one part of the model in turn.

Norms and Games

You can choose the norm on which students need the most work, for your lesson's part of model, and select a game from the appropriate chapter. Note how you will address the norm through discussion by writing in a question you will ask the class. For example, "How does the game show us how to build off of other people's ideas?"

Body of the Lesson

The table in the lesson plan is formatted to enable you to note all the important parts of a lesson:

- How you will open and close the lesson
- The specific tasks in which students will engage
- What teaching moves you will use for each task
- What you think students will do in response to each teaching move

When you are filling in this table, you get to thoroughly think through the arc of the lesson and also indicate timing so that you don't run out of time before you have completed what you planned. You can see that Ms. Cooper filled in possible teaching moves she could use in response to what students may do. She didn't always know what students might do, and she indicated that. Think about the main categories of responses students may have based on early-level understandings. You can't capture every contingency in your plan; that's where improvisation comes in. You will need to use the moves you have learned from this book and from your own experience as students go "off script."

VISUALIZING A LESSON

You may not be able to think of what students may say during the lesson, nor which teaching moves you may use. Talking about your lesson with another person can help you fill in your written lesson plan and serve as a mental rehearsal for the lesson. If you don't have another teacher with whom to do this, you can try it alone. In both situations, you'll be doing what we call visualizing a lesson.

Visualization is designed to help you develop your expertise in the disciplined improvisation required to teach argumentation. By talking through possible paths that students could take in classroom discourse, you prepare to deal with any of those possible paths as they arise in teaching. We have drawn on classic research on how expert teachers plan to develop the process of visualization (Livingston & Borko, 1989).

When you visualize,

1. Start with a written lesson plan that already contains as much detail as you can think of, such as Ms. Cooper's in the first pages of this chapter.

2. Work with a partner, if possible, who can record the results of your visualization.

3. Choose a particularly challenging part of your lesson.

4. Talk through what might happen in chronological order. Let yourself imagine what might happen. Think of the details: your students, your classroom

> Visualizing and written plans work together. The written plan is the foundation for your visualizing and serves as a place to record your ideas.

display, and the tools with which you and your students will be working. Be sure to include both *possible student arguments* and *teaching moves*.

5. Stop and think of possible paths the arguments might take. Explore each of those paths.

Your partner should use active listening to probe what you say for further detail. Active listening requires reflecting back what you say to help you further explain it, to provide you deeper understanding of your own thoughts. It also includes leading you to the next step in the lesson.

You should find that, with practice, you will internalize some of this visualization process and find yourself doing it spontaneously and without always writing it down. As you become more practiced at implementing your teaching moves in the classroom, your planning will become more efficient as you see possibilities and have a ready repertoire of moves to employ.

VIGNETTE: VISUALIZING JUSTIFICATION

In the following vignette, Ms. Cooper visualized just part of a lesson focused on justifying, with the help of Mr. Flores. She chose a part of the lesson that she thought would be challenging for her to teach, in this case, getting students started on justifying. She talked through, in detail, what she imagined would happen, including what she would say, using teaching moves from the earlier chapters in the book, and what the students might say and do in response.

Ms. Cooper and Mr. Flores sat together. Each teacher had the lesson plan Ms. Cooper had written so far on their laptops as a shared digital document. Ms. Cooper was ready to talk aloud, and Mr. Flores was prepared to use active listening and to take bulleted notes on what she said, directly in the lesson plan.

> Ms. Cooper: Yesterday, students made about 10 conjectures. I picked three of them to justify tomorrow. There's one about coordinates on vertical lines, and a similar conjecture about horizontal lines, but that second one is false—they got the coordinates backward. Then I picked the conjecture that I thought was a bridge to where I want to get to—the students using algebra to name the pattern of coordinates in any rectangle. It just states that there is a pattern across the four vertices but doesn't state what that pattern is.
>
> Mr. Flores: So, three conjectures that build to your algebraic conjecture, your "secret mission"? I see them here in your plan.
>
> Ms. Cooper: Yes, and I think I'll start out with the true conjecture, just to ease them into argumentation. They aren't that experienced yet.

(Continued)

(*Continued*)

Mr. Flores: How will you do that?

Ms. Cooper: I'll ask if they think the conjecture is true, and call on someone who does. I don't want to get sidetracked at this point. Besides, it's not that controversial. Oh, I also should show some rectangles on the board, and the coordinates of their vertices—one in each quadrant or one that straddles quadrants, so that they get some experience with negative coordinates. One needs to be a square. One should be "easy"—in the first quadrant, a typical rectangle with different side lengths.

Mr. Flores: What will you do to get them to justify?

Ms. Cooper: I'll just use a standard "why" question. Something like "How do we know this *might* be true?" to get those students who are a little shy to maybe speak up. I bet someone will point to the typical rectangle and say that shows it's true. I'll have to challenge them—maybe ask if one case is enough, or if they can know it is always true based on one example.

Mr. Flores wrote these questions in the lesson plan.

Mr. Flores: OK, what then?

Ms. Cooper: I'm not really sure what they will come up with. But the fact that the sides are parallel to the axes should come in handy—that's how I think of vertical and horizontal, anyway. I think they know that for any point on the y-axis, the x-coordinate is always zero. If they don't, I can ask about that. So if they get that far, then they just have to put those two ideas together.

Mr. Flores, writing down "Students may use parallel and $x = 0$ on y-axis": How do you think they will use that?

Ms. Cooper: Well, there are two ways to define parallel: same distance apart or never intersect. I don't know exactly what they learned last year. I will ask how they define parallel or, if that is tricky, how they know two lines are parallel.

Mr. Flores nodded and wrote down these two questions.

Ms. Cooper: Let's say someone says, "Same distance apart." Then I might have to prompt for "what does that tell us about the coordinates" and even focus them on the meaning of the x-coordinate.

Mr. Flores noted these two teaching moves.

Ms. Cooper: I'm going to assume someone can tell us that the x-coordinate tells you the distance from the axis, and so it has to be the same for all the lines that are parallel to that axis. Well, that's one argument and how it might go.

Mr. Flores: It sounds like you could end up providing a lot of information. How will you make sure students are doing the justifying?

Ms. Cooper: I don't plan to ask every question I told you. As much as possible, I will let students lead the way. I will step in if they get stuck. I can't think of what it would be right now, but they may justify in a different way.

In conclusion, Ms. Cooper thanked Mr. Flores, and he made sure she had access to the changes he had made in their shared file so that she could use them in completing her lesson plan. Ms. Cooper had visualized only about 10 minutes of her lesson—one argument—but it was enough to help her think through the rest of her lesson the next time she sat down to complete her plan. Figure 8.2 shows the portion of the lesson plan that Mr. Flores annotated with what Ms. Cooper said.

FIGURE 8.2 Annotated Portion of Ms. Cooper's Lesson

Task: Students justify three conjectures	Time: 20 minutes Grouping: Whole class
Teaching moves	**Students may**
Introduce the first conjecture. Elicit justification. Show four rectangles on the display: "typical," square, pos and neg coordinates "How do we know this *might* be true?"	Use the typical rectangle as an example, stop there
Do we know if it is always true based on that one example? How can we show it is true for all the rectangles?	Use parallel and x = 0 on y-axis
Ask what parallel means	Say same distance apart. Say never intersect
If needed: What does that tell us about coordinates? What does the x-coordinate mean?	x-coordinate tells distance from y-axis, so has to be the same number at the vertices of vertical lines

As you saw in Chapter 4, when students made this argument, they did not initially refer to parallel. They used as a taken-as-shared fact that points on a vertical line are all the same distance from the y-axis. Ms. Cooper explicitly asked them to tie that to the x-coordinates, which was in her written plan, and she was ready for the argument to end there. But when another student questioned the argument, Lavar used the concept of parallel lines in further explaining his argument to her. Ms. Cooper's planning let her think on her feet about what was essential to the argument and how the ideas connected.

DIGITAL TOOLS

Document-sharing tools are useful when you are visualizing and when you are planning lessons that you want to share with your colleagues. Your school or district may have a set of tools that you are required to use. Even a more traditional word processing program can work, though you must work on the file sequentially, in that case. We provide a digital version of our planning document at resources.corwin.com/mathargumentation, which you can paste into a shared document or file. By using a shared document, Mr. Flores was able to fill in parts of the lesson plan that Ms. Cooper had begun, helping her complete it by adding her own words to the plan.

UPDATING AND SHARING LESSON PLANS

Once you've created a lesson plan, you can use it from year to year. Make it a living document. After you have taught the lesson, go back to your plan and record students' arguments that you didn't expect. Record students' beginning understandings that surfaced during arguments and the understanding the arguments led to. Note the teaching moves you improvised. Think about what you would do differently next time. Through these notes, your teaching will grow along with your plan.

ADVICE ON PLANNING

In this section, we provide advice from both teachers expert in teaching for argumentation and from us, the authors of this book.

Two Teachers and How They Plan

Ms. Acosta works in an urban school with students working at very different levels. She differentiates her plans to include easier and more challenging numbers when students are generating cases. Her plans include the kinds of conjectures she expects students to make at these different levels. For example, when conjecturing about whether even or odd exponents affect the product of exponential expressions with the same base, she expects some students to see a general pattern right away and state it using variables. Other students will be struggling to evaluate 7^2 and so will need extra help in generating cases. Ms. Acosta's struggling students get an extra period of mathematics each day; this is something you can encourage at your school.

> I have different groups of learners in my inclusion classes and classes with more high-performing students. I plan for how each particular group will think about this lesson. Or, if students act this way, I'll ask them this question. But everyone gets to do argumentation.
>
> If students are struggling with computation, I will plan to give them scaffolds that will help—simplified fractions, adding fractions with the same denominator.
>
> —Ms. Acosta

Ms. Elbers teaches in a suburban school with multigrade classes. She tends to work with students in small groups clustered by where they are in the curriculum. As she moves from group to group, she also sees arguments at very different levels. She plans for suggesting specific representations for each group. Ms. Elbers counts on a common planning time with other mathematics teachers in the school to help her keep up with the multiple plans she must make for each class.

> Build argumentation into your daily classroom routine. Sometimes it will be a set of lessons focused on argumentation. Sometimes it will be justifying built into a lesson from the textbook.
>
> I keep a clipboard and camera at my station and record key words I want to make sure they understand. I write down what the students may need to spend more time on or what the next lesson should be.
>
> —Ms. Elbers

Advice From the Authors

We've taken some of the common questions that come up in discussions with teachers on lesson planning and provided our best advice about them.

What's most important for planning for argumentation?

Most importantly, think about how you will get students to talk to each other. It can be easy to step in and do the thinking for students. We are used to thinking of teaching as providing clear explanations of concepts and skills. But your job when teaching for argumentation is to get students to make those explanations, listen carefully to each other, and build off of other students' ideas. Choose moves from this book that facilitate that conversation.

Write down what you can think of in your lesson plan, but know that students will surprise you. Your thorough planning will actually help you in being prepared for the surprises you will encounter.

If you are doing a series of lessons on argumentation, try to plan for all four parts of the model at once, so you get a sense of how argumentation may go over the course of those lessons. Even just a basic outline will help.

When you are doing something new like argumentation, that's when a detailed lesson plan can make a difference. Our form guides you in creating that plan. It's important that it includes both teaching moves and likely student responses.

Do I have to make a complete argumentation lesson plan every day?

In short, no. Some lessons are completely built around one or more stages in our argumentation model, and for those, a lesson plan using our template will help you quite a bit. But for lessons from, say, your textbook, where you anticipate short justifications of conjectures that are already given, you can annotate your regular lesson plan with teaching moves and what you anticipate students may say or do as a result. If, however, you don't regularly make lesson plans, you can use our form to guide you in doing so.

Why should I include standards and learning goals? Isn't one enough?

You'll want to list standards for accountability purposes. Are you addressing the content you need to teach? The learning goals are related, but they are more focused statements of precisely what you want students to understand and be able to do at the conclusion of the lesson. Don't forget "make arguments . . ." in your learning goals and give the topic and type of argument.

What planning steps should I go through?

If your school has a pacing guide or curriculum guide to tell you which standards or which topics you should be addressing for every week or every unit, you can start there. Then find activities that can help you teach those standards. The tasks in this book are resources for many topics, but also look for lessons in which students are asked to explain patterns or explain or justify their answers to problems. Look for prompts in your textbook that call on students to think conceptually, as the concepts are tools for justifications and new concepts are developed through argumentation. Then start thinking about which moves go with which prompts in the lesson. Note some basic moves—especially the "why" questions that you think will be useful. You can also think about the assessments you are accountable to: what sorts of answers students will need to give in open-ended response items. Students' experiences with argumentation in your class should prepare them to address these items.

Should I plan to use whole-class or small-group work for argumentation? Is there a role for individual work?

There are advantages to each mode of class work. Working individually gives students time to see patterns, think about conjectures, or get a start on a justification before participating in a class discussion. You can provide individual work time in concluding so that students can write their own summaries of arguments made by the class. Small-group work will likely be a staple of teaching for argumentation. Small-group work enables students to draw on each other as resources and helps ensure they will talk to each other, which is critical for classroom argumentation. Whole-class argumentation allows you to guide the process a bit more, when the conceptual territory is particularly challenging or when students are just learning to argue together. It also provides time for students to compare arguments made in small groups, for example, drawing connections between representations used. Some teachers provide blank argumentation posters for end-of-unit assessments or weekly check-ins, which is a way of providing students with some practice in making written arguments.

How is planning for each part of the model different?

Generating cases is shaped by the task you pick. Think about the most important cases for students to examine and how many cases you will require them to make. Think about different cases for differentiating instruction.

For conjecturing, reach for what students may think, based on their early understanding of concepts in your lesson. Anticipate that students will surprise you with conjectures that you won't think of in advance.

For justifying, consider the conceptual tools students already have and what the class has established as true already. Think about how some conjectures, once proved true, may be used as part of the justification of a more complex conjecture.

For concluding, think about how to elicit a summary, in logical order, of a whole argument students will have constructed as a class or in small groups.

How do I integrate norms into the lesson plan?

You should play a warm-up game every week for about 10 minutes and write that into your plan. In between, refer to the norms poster—which you may not include in your written plan, but simply use as needed. Because the norms on the poster are tied to stages of argumentation, you can easily see which norms to address.

How do I capture what I know about students' thinking into the lesson plan?

For most lessons, try anticipating three levels of conjecture—a sophisticated conjecture, a middle-level conjecture, and a basic conjecture. Who actually makes those conjectures can surprise you. If you have truly established that students should make bold conjectures, you may hear advanced ideas from students who, for example, struggle with computation.

For justifications, think about what a Level 1 justification would look like and how you could help students reach for Level 2. Consider also what a Level 3 justification would be like and which moves would help students who start at Level 2 to create a more logically complete argument.

How do I plan for contingencies—when things don't go as planned or when students have nothing to say?

That's where you will count on your own improvisation. You may not know how to plan for that, but rather call on the moves you have picked up through the book and doing the Working Together sections. But you can think about the "what" and "why" questions to go to if students don't know what to say at all.

How do I talk to my colleagues about planning?

Even if your department has a common work period, you may not always get to lesson planning because of the other activities on the agenda. Yet common lesson planning can be one of the most important activities you do with other teachers. Try to prioritize lesson planning during designated common time. Try to, at least once a month, sit down with another teacher and visualize part of a lesson.

WORKING TOGETHER

You can begin sharpening your argumentation teaching skills by sharing ideas with your colleagues for planning and visualizing lessons. This work can take place in meetings with your grade-level team, department, or professional learning community. For the meetings, be prepared to share your ideas and build off the ideas of your colleagues, just as you expect students to do.

Exploration and discussion (45 min)

1. In a small group, discuss the advice provided on the lesson planning process. Then think about how you go about planning your lesson. Each person should share his or her lesson planning process with the group. (25 min)

2. In your group, reflect back on your most recent argumentation lesson and discuss the following questions:

   - How did you go about selecting the lesson?
   - Were you able to use the moves in your plan, or did you have to improvise?
   - What did the students do that was unexpected?
   - Explain the changes you will make to improve the lesson. (20 min)

Wrap-up and assignment (15 min)

3. As a whole group, decide how you will plan collaboratively once these sessions are over. (10 min)

4. What are some of the advantages of this type of collaborative activity at your school? (5 min)

 Online Professional Development Guide

For an online Professional Development Guide with activities, games, and lesson planning tools, visit the companion website at

resources.corwin.com/mathargumentation

NOTES

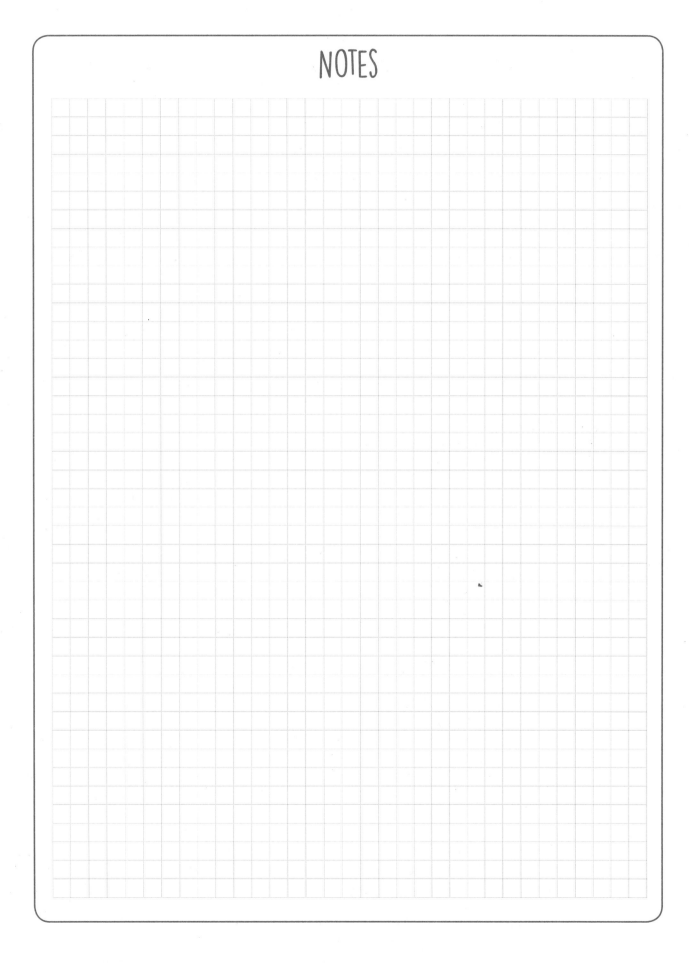

Glossary

Case: A *case* is one number, expression, geometric shape, or other mathematical object that can be used as part of an argument.

Conclusion: *Concluding* means deciding whether a conjecture is true or false based on a justification.

Conjecture: A *conjecture* is a mathematical statement that you think might be true.

Disciplined improvisation: *Disciplined improvisation* is teaching that relies on knowledge of mathematics, argumentation, and teaching moves to support argumentation spontaneously as it happens in the classroom.

Domain: A distinct set of possible values, or a defined set of numbers or shapes used in an argument. For example, a conjecture can have a *domain* of only whole numbers.

Justification: A *justification* is the reason why a conjecture is true or false. A good justification is a connected chain of statements that convinces others of the truth or falsity of a statement; it goes beyond a personal exploration of an idea.

Move: A *move* is the smallest piece of behavior that can be aimed at a purpose. We discuss teaching moves throughout this book as questions you can ask or actions you can take, along with their specific purposes.

Visualization: *Visualization* is a process of imagining how a lesson will go in detail, including imagined student responses.

References

Adams, K. (n.d.). *Story spine*. Retrieved from http://improvencyclopedia.org/games//Story_Spine.html

Adams, K. (2007). *The art of spontaneous theater: How to improvise a full-length play*. New York: Allworth Press.

Aguirre J., Mayfield-Ingram, K., & Martin, D. B. (2013). *The impact of identity in K–8 mathematics: Rethinking equity-based practice*. Reston, VA: National Council of Teachers of Mathematics.

Balacheff, N. (1988). Aspects of proof in pupils' practice of school mathematics. In D. Pimm (Ed.), *Mathematics, teachers and children* (pp. 216–235). London: Hodder & Stoughton.

Ball, D. L., & Bass, H. (2000). Making believe: The collective construction of public knowledge in the elementary classroom. In D. Phillips (Ed.), *Yearbook of the National Society for the Study of Education: Constructivism in education* (pp. 193–224). Chicago: University of Chicago Press.

Boaler, J., & Greeno, J. (2000). Identity, agency, and knowing in mathematics worlds. In J. Boaler (Ed.), *Multiple perspectives in mathematics teaching and learning* (pp. 171–200). Westport, CT: Ablex.

Boaler, J., & Humphreys, C. (2005). *Connecting mathematical ideas: Middle school video cases to support teaching and learning*. Portsmouth, NH: Heinemann.

Campbell, T., Schwarz, C., & Windschitl, M. (2016). What we call misconceptions may be necessary stepping-stones toward making sense of the world. *Science Scope*, *39*(7), 19–24.

Chapin, S., O'Connor, C., & Anderson, N. (2003). *Classroom discussions: Using math talk to help students learn, grades 1–6*. Sausalito, CA: Math Solutions.

Cioe, M., King, S., Ostien, D., Pansa, N., & Staples, M. (2015). Moving students to "the why?" *Mathematics Teaching in the Middle Grades*, *20*(8), 484–491.

Civil, M., & Planas, N. (2004). Participation in the mathematics classroom: Does every student have a voice? *For the Learning of Mathematics*, *24*(1), 7–12.

Cook, G. (2015, July 24). The singular mind of Terry Tao. *New York Times Magazine*. Retrieved from http://nyti.ms/1LAJQH0

De Villiers, M. D. (1999). *Rethinking proof with the Geometer's Sketchpad*. Berkeley, CA: Key Curriculum Press.

Diggles, D. (2004). *Improv for actors*. New York: Allworth Press.

Empson, S. B., & Levi, L. (2011). *Extending children's mathematics: Fractions and decimals*. Portsmouth, NH: Heinemann.

Esmonde, I. (2009). Ideas and identities: Supporting equity in cooperative mathematics learning. *Review of Educational Research*, *79*(2), 1008–1043.

Fennell, F. S., Kobett, B. M., & Wray, J. A. (2017). *The formative 5: Everyday assessment techniques for every math classroom*. Thousand Oaks, CA: Corwin.

Hall, W. (2014). *The playbook: Improv games for performers*. San Francisco, CA: William Hall and Fratelli Bologna.

Harel, G., & Sowder, L. (1998). Students' proof schemes. In E. Dubinsky, A. Schoenfeld, & J. Kaput (Eds.), *Research on collegiate mathematics education* (Vol. 3, pp. 234–283). Providence, RI: American Mathematical Society.

Harel, G., & Sowder, L. (2007). Toward a comprehensive perspective on proof. In F. Lester (Ed.), *Second handbook of research on mathematics teaching and learning*. Reston, VA: National Council of Teachers of Mathematics.

Healy, L., & Hoyles, C. (2000). A study of proof conceptions in algebra. *Journal for Research in Mathematics Education, 31*(4), 396–428.

Inglis, M., Mejia-Ramon, J. P., & Simpson, A. (2007). Modeling mathematical argumentation: The importance of qualification. *Educational Studies in Mathematics, 66*(1), 3–21.

Johnstone, K. (2012). *Impro: Improvisation and the theatre*. Chicago, IL: Routledge.

Knudsen, J., Lara-Meloy, T., Stevens, H., & Rutstein, D. (2014). Advice for mathematical argumentation. *Mathematics Teaching in the Middle School, 19*(8), 495–500.

Knudsen, J., & Shechtman, N. (2017). Professional development that bridges the gap between workshop and classroom through disciplined improvisation. In S. Goldman & Z. Kabayadondo (Eds.), *Taking design thinking to school: How the technology of design can transform teachers, learners, and classrooms*. New York: Routledge.

Knudsen, J., Shechtman, N., Lara-Meloy, T., Stevens, H., & Rutstein, D. (2015). *A professional development program for mathematical argumentation: Bridging from workshop to classroom*. Menlo Park, CA: SRI International.

Knuth, E. J., Choppin, J. M., & Bieda, K. N. (2009). Middle school students' production of mathematical justifications. In D. A. Stylianou, M. L. Blanton, & E. J. Knuth (Eds.), *Teaching and learning proof across the grades: A K–16 perspective* (pp. 153–170). Mahwah, NJ: Lawrence Erlbaum.

Krummheuer, G. (1995). The ethnography of argumentation. In P. Cobb & H. Bauersfeld (Eds.), *The emergence of mathematical meaning: The interaction in classroom cultures* (pp. 229–269). Hillsdale, NJ: Lawrence Erlbaum.

Lakatos, I. (1976). *Proofs and refutations*. Cambridge, UK: Cambridge University Press.

Lannin, J. K., Ellis, A. B., & Elliott, R. (2011). *Developing essential understanding of mathematical reasoning for teaching mathematics in prekindergarten–grade 8*. Reston, VA: National Council of Teachers of Mathematics.

Lara-Meloy, T., & Barros, A. (2000). Base × height: The transformation of a rectangle. *Hands On!, 23*(2), 4–7.

Livingston, C., & Borko, H. (1989). Expert and novice differences in teaching: A cognitive analysis and implications for teacher education. *Journal of Teacher Education, 40*, 36–42.

Lobman, C., & Lundquist, M. (2007). *Unscripted learning: Using improv activities across the K–8 curriculum*. New York: Teachers College Press.

Lockwood, E., Ellis, A. B., Dogan, M. F., Williams, C., & Knuth, E. (2012). A framework for mathematicians' example-related activity when exploring and proving mathematical conjectures. In L. R. Van Zoest, J.-J. Lo, & J. L. Kratky (Eds.), *Proceedings of the 34th annual meeting of the North American chapter of the International Group for the Psychology of Mathematics Education* (pp. 151–158). Kalamazoo: Western Michigan University Press.

Mason, J., Burton, L., & Stacey, K. (1982). *Thinking mathematically*. London: Addison-Wesley.

Moschkovich, J. (2002). A situated and sociocultural perspective on bilingual mathematics learners. *Mathematical Thinking and Learning*, 4(2 & 3), 189–212.

National Council of Teachers of Mathematics. (2000). *Principles and standards for school mathematics*. Reston, VA: Author.

National Council of Teachers of Mathematics. (2014). Procedural fluency in mathematics: A position of the National Council of Teachers of Mathematics. Retrieved from http://www.nctm.org/Standards-and-Positions/Position-Statements/Procedural-Fluency-in-Mathematics/

National Governors Association Center for Best Practices, Council of Chief State School Officers. (2010). *Common core state standards: Mathematics*. Washington, DC: Author.

National Research Council. (2001). *Adding it up: Helping children learn mathematics*. Washington, DC: National Academies Press.

National School Reform Faculty. (2014). ATLAS—Learning from student work protocol. Retrieved from https://www.nsrfharmony.org/system/files/protocols/atlas_lfsw_0.pdf

Partnership for 21st Century Skills. (2008). *21st century skills, education & competitiveness: A resource and policy guide*. Tucson, AZ: Author.

Roschelle, J., Shechtman, N., Tatar, D., Hegedus, S., Hopkins, B., Empson, S., Knudsen, J., & Gallagher, L. P. (2010). Integration of technology, curriculum, and professional development for advancing middle school mathematics: Three large-scale studies. *American Educational Research Journal*, 47(4), 833–878.

Sawyer, R. K. (2004). Creative teaching: Collaborative discussion as disciplined improvisation. *Educational Researcher*, 33(2), 12–20.

Sawyer, R. K. (2011). What makes good teachers great? The artful balance of structure and improvisation. In R. K. Sawyer (Ed.), *Structure and improvisation in creative teaching* (pp. 1–24). New York: Cambridge University Press.

Schifter, D. (2009). Representation-based proof in the elementary grades. In D. A. Stylianou, M. L. Blanton, & E. J. Knuth (Eds.), *Teaching and learning proof across the grades: A K–16 perspective* (pp. 87–101). London: Routledge.

Schmidt, W., Burroughs, N., Zoido, P., & Houang, R. (2015). The role of schooling in perpetuating educational inequality: An international perspective. *Educational Researcher*, 44(7), 371–386.

Shechtman, N., & Knudsen, J. (2011). Bringing out the playful side of mathematics: Using methods from improvisational theater in professional development for urban middle school math teachers. *Play and Culture Studies, 11*, 105–134.

Sinclair, N., Pimm, D., & Skelin, M. (2012). *Developing essential understanding of geometry: Grades 9–12*. Reston, VA: National Council of Teachers of Mathematics.

Stinson, D. W., Jett, C. C., & Williams, B. A. (2013). Counterstories from mathematically successful African American male students: Implications for mathematics teachers and teacher educators. In J. Leonard & D. B. Martin (Eds.), *The brilliance of black children in mathematics: Beyond the numbers and toward new discourse* (pp. 221–245). Charlotte, NC: IAP.

Stylianides, A. J. (2007). Proof and proving in school mathematics. *Journal for Research in Mathematics Education, 38*, 289–321.

Sweller, J. (1994). Cognitive load theory, learning difficulty and instruction design. *Learning and Instruction, 4*, 295–312.

Texas Education Agency. (2012). The Revised Math TEKS (2012): Applying the mathematical process standards. Retrieved from http://texasmathsupportcenter.org/?lesson=1-2-exploring-the-mathematical-process-standards

Thompson, D., & Schultz-Ferrell, K. (2008). *Introduction to reasoning and proof: Grades 6–8*. Portsmouth, NH: Heinemann.

Thurston, W. P. (1998). On proof and progress in mathematics. In T. Tymoczko (Ed.), *New directions in the philosophy of mathematics* (pp. 337–355). Princeton, NJ: Princeton University Press.

Warren, B. (2008). Drama: Using the imagination as a stepping stone for personal growth. In B. Warren (Ed.), *Using the creative arts in therapy and healthcare: A practical introduction* (pp. 118–134). New York: Routledge.

Williams, C., Akinsiku, O., Walkington, C., Cooper, J., Ellis, A., Kalish, C., & Knuth, E. (2011). Understanding students' similarity and typicality judgments in and out of mathematics. In L. R. Wiest & T. Lamberg (Eds.), *Proceedings of the 33rd annual meeting of the North American Chapter of the International Group for the Psychology of Mathematics Education* (pp. 1180–1189). Reno: University of Nevada Press.

Zaslavsky, O., Aricha-Metzer, I., Thoms, M., Sabouri, P., & Brulhardt, A. (2016, April). *Generic use of examples for proving*. Paper presented at Research Presession of the National Council of Teachers of Mathematics, San Francisco.

Index

NOTES

NOTES

NOTES

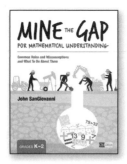 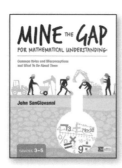 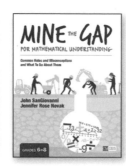

Supporting Teachers, Empowering Learners

The what, when, and how of teaching practices that evidence shows work best for student learning in mathematics.

John Hattie, Douglas Fisher, Nancy Frey, Linda M. Gojak, Sara Delano Moore, William Mellman

Grades K–12, ISBN: 978-1-5063-6294-6
List Price: $36.95, Your Price: $29.95

Move the needle on math instruction with these 5 assessment techniques!

Francis (Skip) Fennell, Beth McCord Kobett, Jonathan A. Wray

Grades K–8, ISBN: 978-1-5063-3750-0
List Price: $30.95, Your Price: $24.95

Students pursue problems they're curious about, not problems they're told to solve!

Gerald Aungst

Grades K–12, ISBN: 978-1-4833-9142-7
List Price: $29.95, Your Price: $23.95

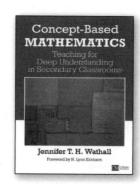

Give students the connections between what they learn and how they do math—and suddenly math makes sense!

Jennifer T. H. Wathall

Grades 6–12, ISBN: 978-1-5063-1494-5
List Price: $32.95, Your Price: $26.95

Differentiation that shifts your instruction and boosts ALL student learning!

Nanci N. Smith

Grades K–5, ISBN: 978-1-5063-4073-9
Grades 6–12, ISBN: 978-1-5063-4074-6
List Price: $36.95, Your Price: $29.95 per book

Everything you need to promote mathematical thinking and learning!

Page Keeley, Cheryl Rose Tobey

Grades K–12
Volume 1, ISBN: 978-1-4129-6812-6
Volume 2, ISBN: 978-1-5063-1139-5
List Price: $36.95, Your Price: $29.95 per book

Corwin educator discount
★ ★ ★
20% OFF
EVERY DAY!
★ ★ ★
Quoted price includes discount.

CM CORWIN MATHEMATICS

N17818

A SAGE Publishing Company

Helping educators make the greatest impact

CORWIN HAS ONE MISSION: to enhance education through intentional professional learning.

We build long-term relationships with our authors, educators, clients, and associations who partner with us to develop and continuously improve the best evidence-based practices that establish and support lifelong learning.